JOHN HUNTER IN 1793.

Aged 65. He died 16th October 1793.

(After the pencil drawing by Sir Nathaniel Dance Holland, Bt.)

Frontispiece.]

MENDERS OF THE MAIMED

THE ANATOMICAL & PHYSIOLOGICAL PRINCIPLES UNDERLYING THE TREATMENT OF INJURIES TO MUSCLES, NERVES, BONES, & JOINTS

BY

ARTHUR KEITH

M.D. (ABDN.), F.R.C.S. (ENG.), LL.D. (ABDN.), F.R.S.

CONSERVATOR OF THE MUSEUM AND HUNTERIAN PROFESSOR
ROYAL COLLEGE OF SURGEONS, ENGLAND

ROBERT E. KRIEGER PUBLISHING COMPANY
HUNTINGTON, NEW YORK
1975

Original Edition 1919
Reprint 1975

Printed and Published by
ROBERT E. KRIEGER PUBLISHING CO., INC.
645 NEW YORK AVENUE
HUNTINGTON, NEW YORK 11743

Originally Published by
HENRY FROWDE – HODDER & STOUGHTON
Oxford University Press, London
Reprinted by Arrangement

INTRODUCTION to the 1975 Reprint

By William LeFanu

Arthur Keith was at the summit of his achievement when he published this remarkable book *Menders of the Maimed* in June 1919 immediately after the end of World War I. The book was and still is remarkable because it combines a masterly anatomical study of bones and joints with a scholarly history of the development of orthopaedics through the nineteenth century in Britain, France and North America. When his book appeared Keith was aged fifty-three and had spent twenty-five years in London as a most successful teacher and public lecturer. *Menders of the Maimed,* as he pointed out in his Autobiography, was rewritten from a score of lectures delivered during the war years at the Royal College of Surgeons. He was throughout his long career an admirable lecturer; his discourses and demonstrations were very carefully prepared, and he shaped them coherently and logically, but spoke in a slightly hesitant manner so that he seemed to be thinking out his conclusions as he went along and persuaded his audience that they were sharing his process of discovery, not merely absorbing ascertained facts.

Keith's own researches had extended into three main fields: human embryology, physical anthropology, and the study of malformations; to each of these disciplines he had made a worthwhile contribution. His textbook *Human Embryology and Morphology,* published in 1902, quickly became a favorite among medical students in Britain; Keith brought it up to date at regular intervals, revising the sixth edition when he was eighty-two years old. His work in anthropology began, while he was a young physician in the Far East, with studies of the muscles of primates, and continued with observations of living human types during travel in Asia, Australia and Europe. Then the comparative evaluation of fossil human remains became the dominant interest of the second half of his life, sparked off as early as 1895 by Eugene Dubois' discovery of 'Pithecanthropus erectus' in Java, though his first influential book in this field, *The Antiquity of Man,* did not appear till 1915; it was a critical survey of opinions then current about the phylogeny of the various prehistoric human types so far discovered. As to his interest in the origin of malformations, this was aroused while preparing material for his teaching courses in embryology. He explored the pathology museums of many hospitals to survey their specimens of the whole range of malformations, but concentrated in his late thirties on the dissection of malformed hearts, an investigation undertaken to support the clinical research of James Mackenzie the eminent heart physician. In the course of this cardiac research Keith discovered the function of the sino-atrial node as the 'pacemaker' of the heart, and published his observation in 1906-07 in collaboration with his pupil Martin Flack. This new discovery proved useful to cardiologists at once, and became of primary value when surgeons began to operate on the heart thirty years later. Though he made his career in teaching and research Keith was fully equipped as a practical physician and surgeon: graduating from the University of Aberdeen, Scotland, he gained experience as intern in a mental hospital, followed by a period in general practice in an English industrial city and a longer spell as a physician in Siam. Then after special studies in anatomy he obtained the higher degree of M.D. from his old University and earned by examination the highest British qualification in surgery, Fellowship of the Royal College of Surgeons of England.

When war broke out in Europe in the summer of 1914 Keith, in his forty-ninth year was too old for active service, but he directed his lecture-courses to support the war effort, first by teaching anatomy to auxiliary first-aid workers, who flocked enthusiastically to learn from him, and then by this course of twenty lectures for the further education of young surgeons faced by the problems presented by war injuries of the extremities. Keith at the same time was forming, with official support, a collection of specimens of war injuries in the Hunterian Museum, which was in all but its name the British National Museum of Medicine, where he was Curator. These war injuries of 1914-18 were new in their severity and had been incurred on a scale never before imagined; but, also for the first time in history, the regular medical staff of the fighting forces was enlarged in proportion to the vast new armies by an organized reserve drawn out of the whole range of the civilian profession from senior consultants to young men just finishing medical school. Keith addressed himself particularly to such young surgeons, and chose to deal with the orthopedic treatment of injuries, because a most urgent and extensive need was the restoration of movement to disabled limbs. Orthopedic surgery was rapidly evolving in response to this demand, and Keith realized that the training and practice which the new specialty required must be based on firm scientific principles or they would be in danger of devising merely empirical techniques of treatment.

When he redrafted these lectures to form this book, he wrote in his preface that it was 'in the main a book of applied anatomy and physiology' showing how the orthopedic surgeon could achieve the most satisfactory results 'by aiding and abetting the natural defensive and recuperative mechanisms of the human body'. Thirty years later, while writing his Autobiography, Keith recalled that 'old and unanswered questions' rose in his mind at this time, whether for instance rest or movement was the better treatment for disabled joints. He had turned, he remembered, to re-read John Hunter's works and John Hilton's *Rest and Pain.* He studied the writing of earlier orthopedic surgeons, particularly the Americans of what he termed 'the New York school', but he also examined the practice of bone-setters and sprain-rubbers usually considered quacks by the medical profession in Britain. He read the available literature on living bone and living osteoblasts. Through this preparatory work he became deeply interested in the men themselves who had made the advances towards modern knowledge and practice, and thus turned himself into a historian as well as a scientist. But on publication Keith's book did not make the impact which he had hoped for it. He had written it in an almost conversational style, quite suitable for a persuasive lecture but certainly not the style which was expected in a serious medical book sixty years ago. On the other hand, 'my publishers dressed it', he wrote, 'in the covers usually given to heavy monographs, and Menders does not belong to that class'. He must have rejoiced in proportionate compensation for his original disappointment when the book was reprinted in 1952, three years before his death at the age of eighty-eight, for he was active and alert to the end.

Now when another quarter century has nearly passed *Menders of the Maimed* appears again for a new generation of orthopedic surgeons. They may ask, perhaps, what this dated survey has to teach them now, since it was written so long ago before the multiple complicated injuries of our mechanized age had become excessively familiar, and more notable before the surgeon was provided with the technical aids which are now assumed as the basic minimum equipment for the preliminary routines of anesthesia and the control of shock. Orthopedic surgery itself has experienced advances in scope and methods which were not even dreamed about when Keith was writing. Yet in fact he offered a clear-sighted exposition of principles which, in his own words, 'underlie the treatment of injuries to muscles, nerves, bones and joints', and these principles do not change with time. Keith's definition is wide enough to include the whole of accident surgery, and he calculated that in writing about men not things he would interest anyone who wished to find out when and by what pioneers victory had been achieved over disabling injuries of many kinds. However by limiting his scope to the treatment of such injuries, as the title which he gave to his book implies, Keith deliberately omitted several equally important branches of orthopedic practice. He barely touched for instance the correction of congenital deformities, which had been the original preoccupation of orthopedists and close to his own earlier interest in malformations. A more general limitation was, as he admitted, that he wrote 'frankly from the British point of view'. German work, not surprisingly in the atmosphere created by World War I, was almost though not entirely ignored, but the most important North American and French contributions to the advance of orthopedics were treated with the detail they deserve, and the brilliant Italian school of Giuliano Vanghetti was not overlooked.

As Keith read the story, advance had been made from three directions: from the bedside in the hospital ward, from the experimental bench in the physiological laboratory, and from the dissecting table in the anatomist's workroom. His aim was to correlate these three strands of discovery, and it was characteristic of his way of looking at history that he saw scientific progress emerging less from thought than from the practical experiments and observations of individual men, whose personalities as well as their intelligence and trained skills had contributed to their successful discoveries; accordingly their careers were worth examining and describing. He believed that this imaginative reliving of former research was the only way to make medical history 'profitable' to writer and reader. From false trails and conclusions later disproved we may learn much about the creative aspects of science; contradictory opinions may each contain elements of value, as in the controversy already mentioned about the desirability of rest or movement.

Keith had grown up in Victorian times and, though a natural gaiety of spirit and some scepticism modified the Scottish Presbyterian influences of his boyhood, he inherited the instinct that a teacher had a duty to make his knowledge as widely and clearly available as possible. Further, he felt in himself the urge to speak out as a popularizer of science in the tradition of Huxley or Faraday whom he called 'the great magician' as both discoverer and popular lecturer. Some of his colleagues thought that he lowered the value of science by writing, as he liked to do, in persuasive conversational language; but for better or worse he kept to his own style, addressing his readers personally like an old-fashioned novelist, in 'an attempt to break the bread of science,' as he put it, 'to the English public which was hungry not for information but for relaxation'. One rhetorical device which Keith customarily employed in his lectures was asking questions and then appearing to work towards the answers by a process of weighing evidence from opposite sides. This device he carried over into his writing, so that in the preface to this book the reader will find himself asked 'What is the rightful place of rest, action, massage, bone-setting? of operative procedures, grafting, suturing, transplantation?' or, looking more deeply into the underlying physiology: 'Is periosteum capable of forming bone?' Instead of answering such questions directly, Keith leads his reader through the conclusions propounded by successive workers, so that reader and writer together work through the earlier researches again. Here, for instance, in Chapter 2 Keith recounts John Hilton's arguments in favor of Rest (1863), while in Chapter 12 we examine Lucas-Championnière's advocacy of 'movement as a means of treatment' twenty years after Hilton, and then go back forty years to Ling's introduction of 'Swedish' gymnastics and therapeutic massage. Keith did not write his history in a stiff chronological schedule, but grouped his chapters to cover a series of different phases of his subject-matter. Thus Chapters 5 to 8 explain the bearing on orthopedics of the great advances made by British and French neurologists in understanding the physiology of muscle, keyed to the concept that 'nerve controls not muscle but movement'. It is worth noticing that the examples Keith uses are not all quoted from famous authorities, but that he brings into his story lesser men whose contributions the reader may be the more ready to explore from their unfamiliarity. Chapters 10 to 13 form another group in which Keith surveyed the varying approaches of workers in different national traditions to orthopedic problems; Chapter 11 will be read with particular interest for its demonstration of Keith's keen appreciation of the early and prolonged American success in orthopedic surgery, to which he returns in Chapter 17 on bone-grafting and transplantation, where F. H. Albee was a pioneer in Keith's own time. The research carried through by Albee's predecessors is discussed in another group of Chapters (14 to 16) where the study of bone-growth is recorded from the experiments of Duhamel de Monceau in France and of John Belchier and John Hunter in England in the eighteenth century through the work of Syme and Goodsir in the 1840s to the conclusions of Ollier and Macewen in the 1880s and 1890s.

Similar to the chapters on bone-growth is the excellent survey in Chapter 19 of the development of knowledge concerning articular cartilage, a problem which had perplexed anatomists at least since the seventeenth century. Keith began his account with the work of Joseph Toynbee who made the first useful research on cartilage in 1838 as a very young man before turning to his main life-work as an otologist. The study of chrondroblasts and fibroblasts is clearly recorded; Keith also as a loyal alumnus of Aberdeen examined the research on cartilage carried out there, before his time at the university, by Peter Redfern and, later, Alexander Ogston. The final chapter on 'Bone-setting ancient and modern' is an unprejudiced history of the subject during two centuries, emphasizing the merits of osteopathic surgery, which in Keith's day was dismissed without consideration by most of the medical profession in Britain.

While dealing with basic physiology Keith had no need to discuss the technical procedures of operative surgery, and in the half-century since he wrote the methods of orthopedics have changed beyond recognition. The introduction of new metals and then of plastic grafts, the innovations in fracture treatment pioneered by Lorenz Böhler in Vienna and Vittorio Putti at Bologna, the recognition of intervertebral disc lesions, the management of paralyzed muscle called for in the great epidemics of poliomyelitis, the effects of deficiency diseases on bone are only a few of the new aspects of orthopedic practice, yet they all vindicate Keith's wisdom in insisting on basic principles which have not, like the techniques of his day, been superseded.

Except for his editorship of the *Journal of Anatomy*, the premier British periodical in his specialty, and his continued attention to all that was being done in embryology, Keith turned away in his later years from strictly medical research. He was active in promoting the spread of scientific knowledge through various societies, by his presidency of the British Association in 1927 and by much controversial writing. His own research concentrated on physical anthropology. Many discoveries of fossil human remains were submitted for his opinion, and he himself investigated many others, notably the Stone Age skeletons from Mount Carmel, which he published in 1939. In his last years he wrote extensively, but without evoking much response, on the correlation of the moral with the physical evolution of man. Certainly his best work was accomplished as a teacher, and *Menders of the Maimed* may claim to be his most generally interesting book, and in spite of its slow start probably the most influential.

PREFACE

MEN of business find it necessary from time to time to take an inventory of the goods they have in stock; occasions arise when medical men must do the same thing and make a survey of the means of treatment at their disposal. That is the case now; surgeons are being called on to restore movement to thousands of men who have been lamed or maimed in war; they find it necessary to re-examine the foundations of their science and practice. In this book I have sought to help them by a re-statement of the principles which underlie the art of Orthopædic Surgery. My chapters are based on a score of lectures given at the Royal College of Surgeons of England in the winter 1917–18 on "The Anatomical and Physiological Principles underlying the Treatment of Injuries to Muscles, Nerves, Bones, and Joints." We all agree, I have presumed, that effective and rational treatment must be based on our knowledge of the structure and mechanism of the human body. Hence, in the main, this is a book of applied anatomy and physiology.

The surgical triumphs won since the discovery of anæsthesia, antisepsis, and asepsis have tended to make the surgeon forget that he is not a master but only a servant; the power to heal lies not in his hands but in the recuperative powers of his patient's body. He can assist the natural powers inherent in all living flesh but he cannot replace them. John Hunter devoted his life to prove that fact; those who read the medical lessons of the war aright realise that we have to return to Hunter's point of view. We can help the sick and the maimed in only one way—by aiding and abetting the natural defensive and recuperative mechanisms of the human body. Our methods must be based on a knowledge of these mechanisms.

I have set about my task in a way which I believe to be new. The first advance in every great curative movement is made by a single mind brooding over facts gleaned by the bedside, the experimental bench, or the dissecting table; these are the critical occasions in the history of Medicine; on such occasions are forged the implements of the surgeon's armamentarium. I have sought to guide the reader to the hospital wards, the physiological laboratories, the dissecting rooms, and private workrooms in which the great advances of orthopædic surgery were made, and to introduce to them the "Menders of the Maimed" as they were in the heyday of life. In only this way, I conceive, can Medical History be written profitably.

I have frankly presented my History from a British point of view. Every great medical movement sweeps over all civilised countries; we are all involved in it. Nothing will please me better than that this self-imposed task should induce colleagues in other countries to write the same history from their national point of view.

The surgical problems presented by the war were new in extent, but not new in kind. The doubts which were revived were old; the questions which urgently demanded an answer were of ancient times. What is the rightful place of Rest, Action, Massage, Bone-setting, Sprain-rubbing, Manipulative Surgery, Electricity, Heat, Cold, Counter-irritation, in our orthopædic armamentarium? What can operative measures do—grafting, suturing, transplantation? Is the periosteum capable of forming bone? To those and many other problems I have sought to provide the materials on which the right answer must be based.

Lastly, although I acknowledge in this preface none of the many who have helped me by name, yet I hope I have not forgotten them in my text.

ARTHUR KEITH.

ROYAL COLLEGE OF SURGEONS, LONDON, W.C.,
12*th April* 1919.

CONTENTS

LIST OF ILLUSTRATIONS

PORTRAITS

FIGURES

MENDERS OF THE MAIMED

CHAPTER I

THE ORTHOPÆDIC PRINCIPLES OF JOHN HUNTER

John Hunter's career may be conveniently divided into the following six stages:

(1) His youth, 1728 to 1748, spent on a farm—Long Calderwood, situated in Lanarkshire, some seven miles to the south of Glasgow.

(2) His "dissecting-room" or Covent Garden stage—lasting from 1748, when he had entered his twenty-first year, until the end of 1760, when he had nearly completed his thirty-second year. In that stage he laid the basis of his knowledge in the dissecting room of his brother William. The dissecting room was situated in the north side of Covent Garden, London.

(3) A short "Army surgeon" stage, extending over 1761, 1762, and the earlier half of 1763, when he saw service in Expeditions to Belle-isle and Portugal during the Seven Years' War.

(4) A waiting or "Golden Square" stage, lasting from 1763 to 1768—from his thirty-fifth to his fortieth year. It was one spent in the search of both knowledge and patients, but was only successful as regards the first. This stage ends with his appointment as surgeon to St. George's Hospital and his removal from Golden Square to 42 Jermyn Street, previously occupied by his brother William.

(5) The "Jermyn Street" stage, marked by prosperity, marriage, lecturing, investigation, and overwork, 1768–1783.

(6) The final or "Leicester Square" stage, when he was recognised by his professional brethren as well as by the public as the leading British surgeon. This stage lasted ten years. Hunter died from sudden heart failure, 16th October 1793.

THERE never was a time when we stood more in need of a ready knowledge of the right kind of anatomy and physiology than now. There are in our hospitals—our repairing sheds—thousands of human machines which have been damaged or maimed in the service of their country—awaiting or undergoing repair. We cannot restore action to any machine, much less to that most complex of all machines, the human body, unless we know the arrangement and working of its parts. Whatever means or practices we adopt and apply for the treatment of these disabled men, there is one condition which we must fulfil—we must base our practice on an intimate and accurate knowledge of the structure and function of the body.

In this as in some of my subsequent chapters, I propose to make excursions into the past, and cull, from the lives of men who devoted themselves to the study of the human body, such facts and experiences as may be of use to us at the present time. There are many reasons why I should begin with John Hunter: his life's work is the common inheritance of the Anglo-Saxon peoples; he was still in the full flight of his career when the United States and Britain parted ways; both countries had set out on their appointed courses when his career came to an abrupt end as the winter session of 1793 began. It is not the stretch of intervening years, however, but the crowded series of revolutionary movements which have swept across the fields of surgery and medicine which make him now seem distant to us. In his closing years came a knowledge of oxygen, of combustion, of respiration, of oxidation of the tissues, of heat and temperature—knowledge which threw much of Hunter's best work into the realm of pure history. Then later, in 1839, began the revolution of the microscope, which gave anatomists and physiologists their atomic theory. The result of that revolution was to make us think in terms of living cells. Hunter always thought in terms of living structures, tissues, and organs. Then, finally, came the great Pasteur-Lister revolution; problems which Hunter had struggled to solve—problems of inflammation, infection, and repair—were finally fixed on a new basis of knowledge, and thus another section of Hunter's life's work lost its currency. The great evolutionary movement left Hunter's work unchanged, for he was an evolutionist; but, fortunately for us, he studied the evolution of function rather than form. For medical men it is function rather than structure which matters. Hence it is that when we have deducted those parts of Hunter's labours which have been displaced by the progress of knowledge, there still remains a vast fund of permanent value, not only for us now, but for generations to come. I propose to pass in review only those parts of his writings which bear on the restoration of action to limbs and joints.

Were I to be asked to cite the most important contribution Hunter ever made to surgery, I would reply that it was his clear recognition of the fact that restoration is effected by powers inherent in the living tissues of the patient; the surgeon can only help recovery by tending these powers. This attitude towards the power of living tissue to conquer disease and to restore function occurs again and again throughout his works. I shall quote only one instance. It is from a passage where Hunter speaks of the practice, followed by certain surgeons, of preventing wounds from scabbing. "This practice," he writes, "arises from the conceit of surgeons who think themselves possessed of powers superior to Nature, and therefore have introduced the habit of making sores of all wounds." You will see that Hunter regarded the surgeon as standing in the same relationship to disease and injury as a gardener does to his plants. He may trim and prune, weed and delve, manure and support, and thus help them to thrive, but the power of growth does not lie in his hands—he can only assist that power. That lesson, I dare think, is needed at all times, and in none more than in the present. A soldier waiting wearily for movement to return to a stiffened joint or limp limb is apt to think that the means of recovery lie in the surgeon's hands, and expects a cure to be worked on him by means which are both active and speedy. It is small wonder if under these circumstances the surgeon may be tempted to become more than Nature's mere assistant. In many modern operations the surgeon's share in effecting a cure is unquestionably great. A stone is removed from the bladder, an enlarged prostate from the pelvis, a calculus from the kidney, gall-stones from the cystic duct, or a fulminating appendix from the abdomen, and thus the conditions necessary for a recovery are at once established. That the surgeon's share is large in such cures is beyond dispute. Nature cannot reappose the fractured ends of a broken bone, nor can she set the parts displaced in a fracture-dislocation of the ankle or the dislocation of a joint. There, again, the surgeon's aid is an essential factor for a good recovery. But in the treatment of chronic ailments and disabilities which follow gunshot wounds, it becomes a much more difficult matter to assess Nature's and the surgeon's share in the recovery. It is in these cases that we have to make certain that the means of treatment which we apply are based on a rational and well-established basis.

To understand how this attitude of Hunter towards the cure of disease became a second nature, you must make his acquaintance during those twelve years he worked in his brother's dissecting room in Covent Garden—from 1748 until 1761. He was an unformed Scottish youth of 20 when he entered the dissecting room, and when he left it as a man of 33, to sail with the military expedition against Belleisle—off the mouth of the Loire—as a staff-surgeon, he was better qualified to serve as a scientific observer and investigator than as an exponent of the practical surgery of the time. It was his fortune during those twelve years in London to be associated with men who studied the human body to learn its living mechanism. He was still more fortunate in studying with men who preferred to dissect their knowledge directly from the actual part than to obtain it at second-hand from the printed page. Above all, he learned then to work, to know how to work, to love work, to become work's most living slave. During those twelve years there were intervals of clinical work at Chelsea and St. Bartholomew's; he had been surgeon-pupil and house-surgeon at St. George's Hospital. But the knowledge which was to give him a foremost place amongst surgeons was gathered not by the bedside but within the walls of the dissecting room. Just glance at the researches he carried out during the dissecting-room period of his life. He worked at the descent of the testicle, and recognised that certain mysterious forces were involved in that operation of Nature; he saw that the functional state of the testicle somehow controlled the act of descent; he saw with curious eye what was merely an everyday observation to others—that the roots of shedding teeth were absorbed—that the extraction of a tooth was followed by the removal of its alveolus, and that such changes could not be effected unless the animal body was provided with a mechanism—a marvellous mechanism—for the removal of useless as well as dead parts. If only he could discover how Nature effected such changes, and could copy her methods, then why should he not use such methods to clear away those innocent and malignant growths which so often threatened life itself? The thread of that inquiry runs all through his laborious career: the removal of sequestra, the burrowing of abscesses towards the surface of the body, and the removal of sloughs. He never did discover that opsonic power, that "consciousness of imperfection" which makes useless parts fall a prey to surrounding healthy parts. It was during those twelve years he unravelled the structure of the placenta. He discovered then that nerves must be of different kinds and serve different functions, otherwise no explanation could be given why branches of the first and fifth cranial nerves terminate within the same area of nasal mucous membrane. He examined the earliest stages in the incubation of the chick, noting the origin of the red-blood islands, the manner in which they fused together and the process by which they invaded the body of the embryo. It was then he commenced to probe the secrets of life itself, testing the peculiar properties of living matter by the manner in which it reacted to cold of varying degrees. He proved that an egg was really a living thing; living matter did not putrefy; life keeps tissues sweet. He had inserted dead matter into abscess cavities to test the alleged digestive action of the pus. Above all, he had made an acquaintance with that lowly form of organised life—the hydra or polyp—and realised that even in its simplest form living matter had the power to move, digest, and feel.

With a knowledge of those earlier years of Hunter's professional life, we are in a position to understand his attitude towards a party of wounded soldiers he met with on his arrival at Belleisle. Four wounded men crossed the British lines after being in hiding for a space of four days; one had received two bullets through his thigh, one was shot in the chest, the third was shot through the knee by a bullet which pierced the lower end of the femur, while another was shot through the shoulder, involving the joint. "When they were brought to the hospital their wounds were dressed superficially and they all got well." That is Hunter's account of their surgical treatment; he gives Nature the credit for their recovery. There was also another case, that of a grenadier of the 30th Regiment, who had been shot in the arm, the bullet passing between the biceps and the humerus. He had been taken prisoner by the French. The arm was fomented and a superficial dressing applied by the French surgeon. About a fortnight afterwards the grenadier escaped and came in with the wounds healed. "There only remained a stiffness in the joint of the elbow, which went off by moving it." Hunter cites those cases to support his contention that in many instances nothing is to be gained by opening and enlarging gunshot wounds. I do not cite Hunter's cases to suggest that the treatment which answered at Belleisle in 1762 is applicable to Flanders in 1917. I cite them because they illustrate Hunter's conviction that Nature is the master surgeon. We also note that he cured stiff joints by movement, and that he preferred voluntary to passive movement.

I propose now to pass on to 1791, when Hunter was in his sixty-second year, and within two years of his sudden end, to see him apply his ripest experience to the treatment of orthopædic cases. In March of that year there came into his wards at St. George's Hospital a man suffering from simple fracture of the right femur, situated three inches below the great trochanter. Splints were applied and the man was rested in bed, but three months later it was found that no union had taken place. On inquiring into the matter Hunter then discovered that his patient had also an ununited fracture of the right humerus of old standing. "It being evident from the circumstances of the arm that there was a natural backwardness in the constitution to form bony union." Hunter directed the man "to walk upon crutches and to press as much on the broken thigh as the state of the parts would admit, with a view to rouse the parts to action, forcing them by a species of necessity to strengthen the limb." To the uninitiated, Hunter's language and reasoning seem obscure, but to those who know the meaning he attached to "action" and "necessity" the principles on which he based his treatment are plain. He recognised that the power of repair was inherent in living tissue; it was an essential property of living matter; it was roused into being by injury—his "stimulus of necessity." If he had coined a learned name from Greek or Latin, as has been done in the cases of other obscure properties of living matter, such as "Chemiotaxis," "Heliotropism," he would have impressed his followers, but in reality made them no wiser. He set the man on crutches to call forth anew, by means of injury, the reaction of repair in the tissues between the fractured ends of the bones; the treatment was successful. "In the course of a fortnight there was an evident firmness in the bone, and in less than two months the patient could walk with the assistance of a stick."

That case was under Hunter's care in St. George's Hospital in 1791. Hunter was interested in the man and allowed him to remain in hospital to pick up strength. During that period he was seized with a complaint of his bowels, which carried him off. "The femur was found to be firmly united by bony union, but the humerus admitted of motion in every direction at the fractured part. . . ." "The arm was carefully dissected to examine the state of the fractured parts, between which there was no callus, but a large bag, filled with a glairy fluid, resembling synovia." A false joint had been formed, and inside that joint lay thirty to forty loose bodies, similar to those which occur in certain kinds of disordered joints. Their presence at once diverted Hunter's mind from the "failure of union" to a subject which had often occupied his attention in former times, the origin of loose cartilaginous bodies in joints. One cannot read and understand Hunter's works unless one knows that in his opinion the film of "coagulable lymph" which united the opposite sides of an incised wound was a living bond. That film or plug of coagulable lymph (fibrin) was, he believed, endowed with a most marvellous property. If it lay between the ends of a broken bone, then it took on "from a species of sympathy" the texture and qualities of bone; if between the cut ends of a tendon, then it assumed a tendinous texture; if against the cut end of a nerve or a vessel, then it became nervous or vascular in texture, as the case might be. If, then, blood were effused into a joint and coagulated on the surface of the articular cartilage, what was more natural than to suppose that it would, as it became organised, assume the nature of cartilage and form the bases of the loose bodies which occur within, and disturb the action of joints? The occurrence of loose bodies within the false joint of the humerus he regarded as a vindication of his theory. Hunter's theory accounted for many appearances, but fifty years later the microscope showed that naked-eye appearances are sometimes deceptive. But no future discovery can ever destroy the utility of a working hypothesis.

We see Hunter at his very best when he comes to speak of cases where disordered action or loss of function is due to damaged muscles. From his first day in the dissecting room in the autumn of 1748 to the fatal visit to St. George's Hospital towards the close of 1793, those mysterious and marvellous power machines of the animal body—muscles—held as in a vice his curiosity and his ingenuity. If it is true—and I think it is true—that a knowledge of muscles is the beginning and the end of all orthopædic treatment, then Hunter merits our attention more than any surgeon. Let us see him brought face to face with a case where muscles are involved. He had called on a lady who suffered from the after-effects of a broken patella. She was totally unable to use the limb and was wheeled about in a chair. "Having spent a whole night considering the probable cause of her loss of power, it appeared to me that this space between the two attachments of the rectus (between its origin and patellar insertion), being much shortened, the utmost degree of its contraction would scarcely be able to straighten itself, much less move the patella and leg also. I advised her to sit as before (with her leg dangling free of the floor), but instead of having the leg moved (passive movement to exercise the quadriceps) to move it herself. This she could not in the least effect at first. I considered, however, . . . if the influence of the mind were freely exerted on the muscle, it would gain this power of contraction." So it proved, the lady gradually obtained the control and use of her limbs.

In this case Hunter's treatment was based on his knowledge of muscular action. He had noticed that a muscle works within definite limits; it can contract under the domination of the will until the shortest limit allowed by the extension or flexion of the joint is reached; it can elongate to the farthest limit permitted by the flexion or the extension of the joint. Hunter observed, however, that under certain circumstances, if a tendon be cut, or a bone broken, the muscle could and did contract beyond the shortest limit of voluntary contraction. Every muscle, he supposed, when contracted beyond its normal limit, passed outside the control of the will. In the case of a broken patella he conceived that the extensor quadriceps was in a stage of extreme contraction and therefore lay outside the influence of the will. But he knew that the will could by repeated effort come to dominate a muscle in all positions of extreme contraction; he knew that a voluntary muscle was the most educable of all structures. It was on these principles that he based his treatment for this particular case. How many muscles are there to-day which need the simple reasoning of Hunter's mind to help them to regain their power of action? By a repeated and judicious exercise of his will a patient may do more to help the recovery of muscular function than can be accomplished by the most complicated of gymnastic machines and of electric batteries.

When Hunter was appointed surgeon to St. George's Hospital in 1768—and thus obtained the first real opportunity of applying his peculiar store of biological knowledge—he was already 40 years of age. In the previous year there happened two notable events in his life: he was elected to the Royal Society and had the misfortune to rupture his tendo Achillis while dancing. It is the latter event which interests us here.

He had been suffering from cramp in the calf of his leg (he was subject to muscular spasms and cramp), and as we have mentioned, he had been dancing. He was not married then. He noted that the rupture was unattended by pain; it was only when repair set in that the tendon became sensitive. He noted, further, that he had lost all voluntary control of muscles yoked to the tendon; by no effort of will could he make the muscles of his calf contract. But he also recognised that when asleep the muscle obtained a power of involuntary contraction. He clearly did not understand why the muscles should pass out of action during consciousness, and assume a power to act during sleep. He knew nothing of the structure and functions of the spinal cord, nor the significance of that kind of action we call reflex. That knowledge came with Marshall Hall. The treatment he applied to cases of ruptured tendo Achillis was the following. He encircled the calf of the leg with a bandage to prevent "involuntary movements," and applied an apparatus during the night "to steady the muscles." After the first day or two following the accident he allowed the patient to walk about, with the heel of the boot somewhat raised on the injured limb, and instructed him to step off the inner side of the heel, as is the manner with those who suffer from flat foot. By these simple means union of the ruptured tendon was secured and a full return of power to the muscles of the calf was effected.

At the time Hunter suffered from this accident he had already acquired two acres of land at Earl's Court, where he established what would be described nowadays as an experimental station. Healing of tendons became a matter of interest to him. Ottley states that he cut the tendons of dogs by a subcutaneous method. Hunter's experiments constitute the earliest records we have of such operations. We know from Everard Home that one of the deer then kept at Earl's Court had a tendo Achillis cut for a double purpose—to keep it from jumping the enclosing fence at Earl's Court, and at the same time to provide information as to the mode in which the breach in a tendon was healed. He found that a space between the separated ends was no bar to union; coagulated lymph, thrown out between the broken ends, could eventually provide a strong bridge across a wide gap. After Hunter's death it was found that the cicatricial bond in his tendo Achillis had become calcified.

We now come to another side of Hunter's teaching regarding the nature and action of muscles which has an important application to present-day conditions. We may call this part of his teaching, in order to distinguish it, the principle of the "minimal load." Let me explain by use of an example what is meant by that expression. Let us take the case of a labourer who has just recovered from typhoid fever. We want to test his strength. We ask him to lift and carry a sack containing two cwt. of coal. We should be very foolish if, when we find he cannot move it, we conclude that he has no strength. Yet that is how the strength of weakened and disabled muscles is being tested now in many instances. Let us take the case of the deltoid muscle. If we start it into action as the arm hangs by the side, it has to move its maximum load at the very start, it has then to lift the weight of the arm when its purchase is at a minimum. The deltoid may, when so tested, appear absolutely paralysed, yet if the disabled man lies on his back with his arm half abducted, and its weight supported on a board, and we then apply the principle of the "minimal load," it may be found that the muscle, far from being paralysed, has a very considerable power of action.

It is characteristic of Hunter that he never did halt or bend to point out the immediate practical application of his observations and "principles." He gives us, as it were, the stick and the knife, but expects that we ourselves will do the whittling. That is why Hunter is a prince to the thinking surgeon and only a babbler to the merely practical one. He tells you the position of a joint which gives a muscle its greatest power; he points out that the weight of a limb is a heavy burden; he recognises that muscles may be weakened from many causes; he insists on the mastery which the will can obtain over them, but nowhere does he formulate these observations into a system to be applied to the treatment of disabled muscles. Indeed, I must own that without the aid of my friend Dr. W. Colin MacKenzie I should never have realised the practical bearing of many of Hunter's facts. Dr. MacKenzie makes no claim to be the discoverer of the "minimal load" treatment of disabled muscles, but I am certain that no one has realised its practical importance more than he, and no one has realised and applied the right methods to the restoration of disabled muscles with a greater degree of skill. When you have discovered that a muscle loaded to a minimal degree can exhibit a spark of movement, then you have to bring the will to bear upon that spark. As it grows you will be able gradually to increase the load and thus nurse what is often regarded as a useless structure back to a fair or full degree of vigour. Dr. MacKenzie's methods, as revealed by his published results, bespeak infinite patience and perseverance on the part of both surgeon and patient, and these are qualities which are hard to obtain in either surgeon or patient. By citing this instance I again bring you back to the bed-rock of Hunterian teaching: the only rational means of treatment are those which are based on the natural recuperative powers of the body.

One other aspect of Hunter's teaching has a bearing on the treatment of disordered or paralysed muscles. He was not the first to realise that it is impossible for a muscle or a set of muscles to contract unless their antagonists are called into action at the same time. He learned that doctrine as a student from Winslow's excellent text-book of Anatomy; but he did realise more than anyone who went before him that the relaxation of the extensor or antagonist muscle was just as important, as real, and as vital an act as the contraction of the flexor or prime-moving muscle. He never could determine whether the relaxation of the antagonist was due to a voluntary or an involuntary stimulus. That doubt, we shall see, has been resolved in recent times. The surgeon must always think in terms not of single, but of linked or opposed muscle groups.

All through his professional career Hunter was ever seeking to obtain a more intimate knowledge of how injured or diseased parts of the body were healed. His aim was to supply the conditions which were favourable for the recovery of health and to withdraw those which were unfavourable; more than that he believed no surgeon could do. Every case he attended, whether the issue was a success or a failure, led him to readjust his stock of knowledge in the light of the experience which it had yielded him. Hence it was that from year to year his art ripened. "Never ask me," said he to a pupil, "what I have said or written, but ask me what my present opinions are, and I will tell you." Sir Astley Cooper, when a medical student, reminded him that in a previous year he had expressed a different opinion. "Very likely I did," replied Hunter, "I hope I grow wiser every year." In the years just previous to his death he prepared for publication a work [1] into which he compressed the repeated and tested experience of thirty-two restless, crowded but observant years. We are to look for a moment at the means which the experience of a lifetime led Hunter to adopt in the treatment of "Diseased and Wounded Joints respecting their Motion." [2] We are to try and ascertain what Hunter would do if he were now recalled to take his place in one of our modern orthopædic hospitals. We know with certainty that before he would attempt to restore movement to a stiffened joint or limb—that joint or limb must first be free from inflammation—from acute disease. He held that "nothing can promote contraction of a joint so much as motion before the disease is removed. . . . When all inflammation is gone off, and healing has begun, a little motion, and frequently repeated," so Hunter thought, "is necessary to prevent healing taking place with the parts fixed in one position." He preferred voluntary to passive movements—if the muscles still retained any power of volitional contraction. He was well aware that inflammation must be followed by a state of stiffness, for there could be no inflammation without adhesions being formed in the tissues affected. "But the parts will stretch and the motions again

[1] *A Treatise on the Blood, Inflammation, and Gunshot Wounds.* Edited by Everard Home. 1794.

[2] See vol. i. p. 581 (Palmer's edition of *Collected Works*).

become free by gradual motion and by friction." Hunter, then, would treat our modern cases of stiffened limbs and joints by instituting voluntary or passive movements and by the application of massage. "In healing of wounds by granulations" he anticipated that the rigidity would be greater, particularly so "if there has been a loss of substance from mortification and gunshot wounds." In such cases, and they are the commoner kind in our modern military practice, "more attention is requisite. If the inflammation has been violent, or if the disease period of healing has been greatly prolonged, active or voluntary movement on the part of the patient may be ineffective owing to the extreme weakness of the muscles which act on the disabled joint." In such cases we find that Hunter would resort to one of three methods of aiding the weakened muscles to move the segments of a limb. (1) The hands of "another person" may come to their aid; (2) the limb may be placed in such a position that gravity will aid the weakened set of muscles. If the knee is fixed in a flexed position and the quadriceps extensor is weak, then by placing the patient face downwards on his bed, the extensor muscles are aided in their efforts by the weight of the leg and foot. (3) A weakened group of muscles may be aided by loading their opponents as by placing a weight in the hand to aid the triceps extensor of the elbow in overcoming the contracted flexors of that joint.

Hunter regarded, and rightly regarded, muscles as the most "educable" of all the structures of the body. Clearly it was his opinion that a surgeon could not undertake their management and treatment, unless he looked at them with the eye of a psychologist. Let me take one of his examples—a case where the muscles of the calf were in a state of over-contraction and permanently shortened—no matter what the cause of that shortening may have been. His directions for treatment are that the calf muscles must be encouraged to relax and must be prevented from over-contracting. "In this case we should give support to the heel in a degree. By this means the patient, convinced of his ability to bring his heel as low as the support requires, will cease voluntarily to contract the gastrocnemius and soleus muscles and they by degrees will acquire the property of relaxing themselves, which, if the heel was entirely supported at first, they would have no occasion for, and while altogether unsupported they never could." From the instance just cited it will be seen that Hunter regarded muscles as semi-conscious living things which had to be coaxed and humoured back to health. Another instance will show the importance he attached to the mastery of muscles by the brain. "Suppose the elbow is contracted and I want to put the flexor muscles into an elongated state, neither the muscles nor mind being able to do this. The hand apparently is not heavy enough (to overcome the resistance of the shortened flexors of the elbow-joint), but suppose I should then put a little weight into the hand, in order to assist them a little. Then the mind will be fixed on the support of the weight; the flexors will support it, being influenced by the mind. Thus we shall do harm *by increasing the power of the flexors.* But if the mind is fixed on the extensors (on the extension of the elbow-joint), and not at all on the support of the weight, then the weight will be serviceable by extending the flexors while contracting the extensors." Great artists are said to mix their paints with brains; examples, such as I have given, show that the great surgeon makes brains the chief ingredient of his prescription.

Hunter was keenly interested in the wasting of muscles. He saw that when a joint or part became the seat of injury or of disease, the muscles which acted on the joint became rapidly reduced in size and strength. He noted that if the hip-joint were affected, wasting did not proceed beyond the knee, or if the shoulder were the seat of disease, beyond the elbow. He knew that the wasting was not due to want of use, for the opposite sound limb, which was used as little as the diseased limb, showed no corresponding degree of muscular atrophy. Hunter supposed that the muscles wasted from "a species of sympathy" which linked them with the joint on which they act. Nowadays we have changed the phrase, but by "reflex action" we mean exactly what Hunter meant by "a species of sympathy."

He had a clear conception of the wreckage left within a part when the storm of inflammation had subsided. During the storm "coagulable lymph"—our "inflammatory exudate" —had been thrown out in all the "cells" or spaces which permeate and surround every particle of living tissue and moving structure. Hunter had studied the transformation of "coagulable lymph" into adhesions—adhesions which bound together parts which were meant to glide and slide when limbs are moved. Stiffness was a necessary result of inflammation. We have seen that he relied on movements—voluntary in the first place, and passive in a minor degree—to elongate such adhesions, little by little. We have seen he also used friction. "Friction with vinegar, lees of wine, spirits of wine have been recommended. Whether these last remedies serve any useful purposes or not, I am undetermined." "If pressure is used as a stimulant, it must be gentle; if violent it will irritate and produce vesication, as we see in bed-ridden people. Pressure produces increase of action, by first impeding action, the impediment acting as a stimulus on the part. Friction is more powerful, more active and immediate than pressure; it rouses the parts to action and excites a warmth in them. Stimulating or irritating lotions render friction more powerful in torpid parts. Fibres, by frequent rubbing, are stretched and lengthen."

More than anyone, Hunter tried to discover the effects of heat and cold when applied locally to assist healing. "Water, as it conducts quicker, is a better vehicle for the application of heat than air." Tissues which had been weakened by inflammation he supplied with warmth and shelter as they battled for restoration. Such tissues are weak and have lost their normal power of responding by an increase of action of metabolic exchange—when exposed to the depressing effects of cold. In the early stages of treatment warmth was regarded by Hunter as a first necessity. In later stages Hunter had proved by experiment that cold could act as a tissue stimulant —cold applied in an intermittent way. As to electricity, his statement was necessarily guarded by its novelty as a means of treatment. "Electricity," he states, "is the only application, perhaps, which can be said to penetrate, and its influence is very extensive." But as to the exact nature of that influence we know to-day little more than Hunter did in 1793.

We have seen that the means which Hunter applied for the restoration of movement to limbs which had been stiffened by injury or disease was a reasoned system of mechanical therapeutics. His was a system of mechanico-therapy grounded on a knowledge of the powers of the human body. He was persuaded that the power to heal was a property of living tissue, but the power to recover function—the function of joints and muscles—was a property of the patient's will and brain. His opinion was that it is the surgeon's business to direct, encourage, and interest that will and brain. Without the patient's active assistance the surgeon's best efforts are in vain. Our modern experience has shown that Hunter was right. Surgeons who have accepted the principles which he taught and have sought to restore lost or lessened movements by giving the men congenial tasks that lead to voluntary exercise of weakened muscles have succeeded best. If we had followed Hunter we should never have made the mistake of supposing that elaborate batteries of gymnastic machines could take the place of the thinking brain of the surgeon and the willing response of the disabled soldier.

JOHN HILTON,
About the year 1867, when he was 60 years of age.

CHAPTER II

JOHN HILTON'S PRINCIPLES OF TREATMENT

John Hilton was born in 1807 near the rural village of Castle Hedingham, situated in the northern part of the county of Essex, some fifty miles to the north-east of London. Of his people and of his boyhood we are quite ignorant. His life may be marked out into the following stages:

(1) His boyhood from 1807 to 1824, when he became a pupil at Guy's Hospital at the age of 17. Before becoming a medical student he attended the grammar school at Chelmsford, and was for some time in France, at a school in Boulogne.

(2) A "dissecting-room" stage, extending from 1824, when he had entered his eighteenth year, until 1845, when at the age of 38 he became assistant-surgeon to Guy's Hospital and lecturer on Anatomy. In this stage of his life he lived quietly in the suburbs and earned his living by demonstrating anatomy and physiology at Guy's Hospital, and also performing examinations in the post-mortem room. It was in this phase that he carried out his observations on the skull, arteries, and nerves, and laid the foundation of his knowledge of the human body.

(3) A "surgeon-anatomist" stage, extending from 1845 to 1853,

when he lectured on anatomy and served as surgeon at Guy's Hospital. He took up his residence in the City in New Broad Street—where Sir Astley Cooper had practised.

(4) A "Rest and Pain" period which covers his successful professional life from 1853, when he was 46, until 1878, when he died from cancer of the stomach at the age of 71. The last decade of his professional career covers the opening period of Listerism.

A period of thirty-one years intervenes between Hunter's death and the entrance of John Hilton as a student at Guy's Hospital. In that interval medical schools had been established within the walls of the great Metropolitan hospitals. In 1824 Guy's and St. Thomas's Hospitals stood side by side in the Borough near the south end of London Bridge. In Hilton's time the schools of the two hospitals became separated (1836). Through Henry Cline and Astley Cooper, Hunterian teaching was well known at St. Thomas's in Hilton's student days, for both had been pupils of Hunter. Hilton differed altogether from Hunter in that he had not the genius for devising experiments to extend knowledge. But he resembled Hunter in seeking to fathom the meaning of experiments which Nature made for him.

IN the spring of 1862 John Hilton completed in the theatre of the Royal College of Surgeons a series of lectures, which, when published in book form under the title of *Rest and Pain*, won an enduring place in the minds and hearts of medical men. *Rest and Pain* is a title with more than a touch of romance in it; it prepares you to expect the finely chiselled face of a poet; but the man who wove those vivid and incisive lectures from material freshly gathered at the bedside, had nothing of the poet about him. His was a flat, rotund, plainly moulded, robust and honest face. His contemporaries tell us that he had the outward appearance of a dapper, prosperous City man. There was nothing at all of the student's cast of countenance about this man, whose skilful practice we are to follow. But there is much to suggest a taste in dress. He liked a waistcoat with a decisive pattern—one which was linked from pocket to pocket with a heavy gold chain and showed an ample shirt front. There is no trace in him of the Society man; he cold-shouldered all the world, save the few he selected for associates and friends. He could be, and often was, overbearing, rude, and mordantly sarcastic. But we shall see that he faced adversity with a brave heart, asking no favour, and having no fear of any man; knowing well that he had that to sell which the world must buy not at its, but at his price.

Having thus introduced John Hilton when he had established for himself a sure place in British surgery, I propose now to turn to his earlier years, and watch his growth in knowledge. Hunter served the apprenticeship to his life's career in the dissecting room; so did John Hilton. Hunter's dissecting room was in Covent Garden; Hilton's lay in the Borough, beyond the south end of London Bridge. In the year 1824, when Hilton, a lad of 17, came from Chelmsford Grammar School to commence his medical studies, the twin hospitals of St. Thomas's and Guy's were cheek-by-jowl in the Borough. They had a common dissecting room when Hilton attached himself to Guy's. We know that he was born at Castle Hedingham, in the upper reaches of the Colne, in North Essex—as pure a Saxon settlement as is to be found anywhere in England. He was a Saxon in build of body and in frame of mind. As to his people, I can discover nothing, but they were apparently not well off, for Hilton could not afford to pay the surgeon-apprentice's fee which would have secured for him the privileges of climbing to the highest position in the hospital. Hunter had been dead more than thirty years when Hilton entered the Borough in 1824, but his name and teaching were kept green there by two ardent Hunterian pupils and disciples, Henry Cline and Sir Astley Cooper. Hilton could begin where Hunter left off.

In 1828, at the age of 21, having taken the diploma of the College of Surgeons, he entered the dissecting room of Guy's Hospital as demonstrator of Anatomy, and there, and in the post-mortem room, he was destined to abide and work for seventeen long years when, in 1845, at the ripe age of 38, he was appointed assistant-surgeon to the hospital and lecturer on Anatomy. Like Hunter, he served his apprenticeship to surgery in the dissecting room. Only it was a different apprenticeship. There was a home in the suburbs, but no experimental station, no sedulous cross-examination of freezing eggs, incubating chicks, or primitive hydra to yield the secrets of life. Certain things had happened since Hunter's death which influenced the men of Hilton's time. Archdeacon Paley had cast a spell on the marvellous mechanisms of the human body. It became the chief duty of the anatomist to discover the utility and design of every one of its many structures. Hence we find Hilton in the dissecting room seeking continuously for an explanation of the arrangement and use of its various parts. He turned his attention to the distribution of arterial trunks, and sought for a functional, or—as his contemporaries would have phrased it—a teleological explanation. For him the internal maxillary became the "artery of mastication"; it was designed and laid down to supply all the parts concerned with the jaws and the muscles which moved them. Its large middle meningeal branch, far from proving his thesis to be wrong, proved its truth, for did it not supply that temporal area of the skull which provided a firm basis for the largest of masticating muscles—the Temporal? On the same basis he explained the distribution of the branches of the subclavian arteries; they were "respiratory vessels," destined to supply muscles and bones concerned in the act of respiration. The cœliac axis was the "artery of digestion"; it supplied the organs immediately concerned in that act. We see him hot foot in the application of the doctrine of utilitarianism to the structure of the human body.

Certain events which occurred just before and just after he became demonstrator at Guy's in 1828 turned his attention to the manner in which nerves are distributed; these were the discoveries of Charles Bell and of Marshall Hall. On becoming a medical student, Hilton could not have escaped the glamour of the doctrine which Charles Bell was still preaching from the precincts of Piccadilly; for the first time a utilitarian explanation of nerve roots had been given; for the first time it was shown that the nerves going to muscles were compound in nature; they contained fibres which carried messages from the brain to the muscles, and also others which carried messages from the muscles to the brain. It was late in 1832, just as Hilton was commencing his fifth year in the dissecting room, that Marshall Hall, working in a laboratory extemporised in his Bloomsbury home, announced to an incredulous world his discovery of a "reflex"—or, as he preferred to call it, the "excito-motor"—function in the spinal cord. We shall see that Hilton never did understand the nature and full significance of Marshall Hall's discovery, but nevertheless it bent him towards the investigation of nerve distribution. Hence we find him dissecting out the exact distribution of the laryngeal nerves; he shows that the nerve which furnishes the mucous membrane of the larynx with sensory fibres also supplies its muscles with motor fibres. For him that observation became the basis of a law—the nerve which supplied the muscles and controlled the movements of a part also served the skin and other sensory surfaces which were connected with that part. For him the nerve trunk became the functional element or unit; he preferred to work on what seemed a solid anatomical basis rather than to trust himself to the intangible "excito-motor" theory of Marshall Hall. He applied himself to the investigation of the nerve supply of joints. In the hip he observed that the three nerve trunks—anterior crural, obturator, and sciatic—which supply the muscles which act on the joint, also send sentinel branches to the joint itself. By such an arrangement he believed a consensus of action was obtained; the articular sentinels kept their muscular colleagues informed as to the condition of parts within the joint at all stages of a movement. In the knee-joint he found further confirmation of his belief; the obturator nerve sent a twig to the knee-joint because the gracilis—an obturator or adductor muscle—acted on that joint. For the same reason the nerves which supplied the extensors and flexors of the knee also supplied the joint. At the knee-joint he was able to go a step further. The nerves which supplied a joint also supplied the sentient skin which enclosed and protected the joint.

Hilton's application of this theory to the knee-joint presented him with a difficulty; the internal saphenous nerve, which furnishes a large branch to the skin of the knee, is distributed chiefly in the leg and foot. A nerve which observed his theory should not do a thing like that. Hilton's was an eminently logical mind, and this, which seemed an exception to his law of "consensual distribution," was turned to its support in the following manner. The sartorius muscle acted not only on the knee-joint, but through its fascial insertion on the whole of the inner side of the leg. It was necessary, he conceived, that a nerve of a muscle group should supply the skin over its entire area of action, and hence the extension of the distribution of the internal saphenous nerve to the inner side of the leg. He supported his contention by an appeal to the upper extremity; the posterior circumflex —the nerve of the shoulder-joint—sent cutaneous branches beyond the deltoid, because that muscle had an insertion to the fascia of the arm; the musculo-cutaneous of the arm sent a branch to the forearm because the biceps, by its fascial insertion, acted on the forearm. In the dissecting room—and afterwards in his surgical lectures—we shall often find him straining a hypothesis far beyond its breaking-point. But, right or wrong, it was just because he tried to explain appearances and kept his eye ever on the alert, that he became a power in British surgery. In 1839, while still engaged in these dissecting-room researches, he was elected to the Royal Society at the age of 32. In his nomination form he is described as "a gentleman much attached to the science of Anatomy and Physiology."

We are now to follow Hilton when he is launched in the field of surgery, at the ripe age of 38, and watch him apply, in the treatment of joints, muscles, and nerves, the knowledge he had gained during the seventeen years of continuous labour in the dissecting and post-mortem rooms. In some respects he is to out-Hunter Hunter. He never leaves us a moment in doubt as to the rôle the surgeon is to play in the treatment of disease. He must be nothing more, and nothing less, than Nature's humble assistant. At the end of his fourteenth lecture he summed up his philosophy in these words:

"By regarding this subject of physiological and mechanical rest in what I conceive to be its proper professional light, the surgeon will be compelled to admit that *he has no power to repair directly* any injury. It will induce him to acknowledge, in all humility, that it is the prerogative of Nature alone to repair the waste of any structure. He will thus realise that his chief duty consists in ascertaining and removing those impediments which obstruct the reparative process or thwart the effort of Nature, and thus enable her to restore the parts to their normal condition." In such an instance we see that Hilton had the art to clothe in flowing lines certain thoughts which, although familiar to Hunter, found from him but a laboured utterance.

Nor can I resist quoting here another instance from his third lecture, which shows us not only his opinion of the surgeon's place in the temple of Æsculapius, but gives us a glimpse of Hilton's humanity and his command of idiomatic English, learned while a boy at Castle Hedingham.

"A surgeon from the country lately came to my house with a patient. He said: 'I want to consult you about a young lady who has a diseased toe.' With her was a relative, an elderly gentleman, a very kind-hearted man, who thinks himself a good surgeon, and goes about doctoring people, sometimes doing harm and sometimes perhaps a great deal of good. He is very fond of animals, and has a number of pets. After I had examined, with the surgeon, the lady's toe, the elderly gentleman said: 'Well, Mr. Hilton, what are you going to do to cure this young friend of mine?' I said: 'I think I shall put a splint on the foot and keep the toe quiet, attend to her general health, and Nature, in all probability, will do the rest.' I then said to him: 'What led you to adopt the occupation of a philanthropic surgeon?' 'Well, Mr. Hilton,' he replied, 'I will tell you. Some years ago I caught a live mouse in a trap. . . . Then I thought to myself, "This mouse must have had difficulties in the treatment of its injuries; and——" interrupting his story, he said, 'I hope you won't be offended at what I am going to say?' 'No,' said I, 'not in the least.' 'Well,' he continued, 'I said to myself, "Surely this mouse, although it is cured, never had a physician or a surgeon!" I agree with you, Mr. Hilton, that Nature is a very valuable surgeon.'"

The vivacious passage I have cited makes it very clear that Hilton considered the surgeon's first duty was to give the injured parts rest. It was by means of rest that he could best help nature. The therapeutic value of rest has been recognised by surgeons of all periods. It constituted the fundamental principle of Hunter's practice, but while Hunter gave it a place at the end of a paragraph, Hilton made it the heading for a new chapter. He used his knowledge of the structure and function of the body as a means to discover how parts could be rested. He began by applying his dissecting-room investigations to the treatment of inflamed joints. "Why is an inflamed joint fixed and flexed?" he asks. His answer is that "the irritated or inflamed condition of its interior (say the knee-joint) involves all the articular nerves, excites a corresponding condition of irritation in the same nerve trunks which supply its extensor and flexor muscles. The flexors by reason of their superior strength compel the limb to obey *them*, and so force the joint into a *flexed position*. . . . The muscles, indeed, appear to be told, through the medium of the nerves of the interior of the joint, that its articular structures are over-tasked; the antagonistic muscular forces of the joint being thus *involuntarily excited*, the joint is at once rendered rigid and stiff for the purpose of keeping it at rest. . . . The flexors act unceasingly day and night, apparently without rest, and especially declare their mischievous assiduity by the wakeful slumbers and disturbed sleep of the patient." The surgeon's first duty was, therefore, to give the exhausted muscles rest by removing the source of their disturbed action—the irritation or stimulus which arose within the diseased joint at each movement. That could be done in only one way; if we would set the muscles which act on the knee-joint at rest, then we must prevent the articular nerves of the joint from being irritated by the pressure and friction which result from every movement—particularly involuntary movements. There was only one way—rest. But not the rest given by a general instruction to patient, guardians, or nurse—as Hunter gave it; not rest in homœopathic or intermittent doses, but a long, continuous course,

The methods used to secure such rest must be founded on a sure anatomical and physiological basis, prescribed with precision and carried out meticulously until nature had effected a cure. The application of splints was his chief means of securing physiological rest of joints. We shall have occasion later to allude to the designs he employed; his service to surgery lies not in the forms of splints he recommended—but in his insistence on their unremitted application.

In passing at once to Hilton's chief work—his application of the principle of rest to the treatment of diseased and injured joints, I have failed to do justice to the solid basis of observation on which he based his practice. He had noted, as many a surgeon had done before him, that spasm departed from the orbicularis palpebrarum when the irritant, whatever form it might assume, was removed from the eye; that the removal of a fish-bone from the throat set the pharyngeal muscles at rest; and that the extraction of a calculus relieved spasmodic contractures of the bladder. The surgeon's first duty was to discover the site and nature of the irritant; to obtain rest, the irritant had to be removed.

He construed the word "rest" in a wide sense. You will recall the case of a "man who had received a blow on the chest from a fall upon the part . . . I could find no fracture of the ribs; but I observed that the patient had a most worrying wife. I suggested to the physician that the sole cause of the pain was, in all probability, produced by the patient constantly moving the injured or bruised soft parts by using his chest and lungs in speaking. All I recommended was that he should hold his tongue and have his chest bandaged. I requested that his wife should not say a word to him. From that time he got quickly well by local rest."

The instance shows us that he used the word "irritant" as well as "rest" in a wide sense. He considered that the false membrane which forms on an acutely inflamed mucous surface, the scab which spreads over an open wound, the exudates which appear on the surface of inflamed serous membranes, were Nature's means of protecting these parts from irritation, and thus providing them with rest. It was to give rest that he excised painful cicatrices and cut the nerves which supplied the painful points of irritable ulcers. He rested the heart by confining the patient to bed, and elevated the extremities to rest their vessels. He relieved tension to give rest—the tension of abscess, of stretched nerves and muscles. Occasionally he resorted to tenotomy to give a spasmodic muscle rest. One series of cases which he treated by rest for the relief of tension of nerves is of particular interest to us. They are cases which Hilton described as "cervical exostosis," but which we now know are due to the pressure of the lowest trunk of the brachial plexus against the uppermost rib—usually one attached to the seventh cervical vertebra. Hilton treated such cases by rest—rest in the recumbent position, which relieved the suspensory muscles of the shoulders and lifted the nerve trunk from the point at which they pressed against the cervical rib. Whether we agree with Hilton in all his interpretations of Nature's intention or not, we must admit that he sought for and obtained a rational basis for his means of treatment—one founded on the facts of anatomy, physiology, and pathology.

In his scheme of therapeutics, action finds no part—neither active movements or passive movements are considered suitable means of combating disorders of joints or muscles. "No counter-irritation, none of the old horse-doctoring style of treatment, no setons, issues, or painful applications of any kind," is his verdict as regards ancient and widely established practices.

Hilton condemned the application of blisters, counter-irritation, and stimulating lotions not on the ground that his experience had shown their effect to be bad, but because they offended against his first principle—rest. Such applications, he believed, prevented a disordered part from having rest—the condition essential for recovery. On the other hand, he favoured the application of sedatives; when applied to the point or source of irritation they gave rest not only to the diseased part but to the whole system. The instance I am to cite is drawn, not from his practice in joint disease, but from another class of case—one which he treated with great success—namely, disease of the rectum. He applied his principle of rest to surgical disorders of all parts of the body. "To show the relative value of this soothing treatment, as compared with that of local irritants, let me mention another case. A gentleman came to me on the 17th of September 1861. I well recollect the day, because I was cold and shivering from an attack of ague caught in Holland; I was angry, ill-tempered, and felt very uncomfortable. The patient was between sixty and seventy years of age. He told me he had suffered much pain in his rectum, and that he had been under the conjoined care of two surgeons, who had assured him that he had not any cancer, but they could see and feel an ulcer in his gut, to which they had applied, in the form of injections, solutions of nitrate of silver, sulphate of zinc, sulphate of copper, and some preparations of lead. He added, 'I must honestly tell you that although I have taken an immense quantity of medicine, I am a great deal worse than when I went to them.' I proposed to examine his rectum. 'No,' he said, 'you must not examine me; I won't be examined any more. I have suffered so much already from that speculum.' I said, 'You are very foolish; I cannot tell with certainty what is the matter till I have examined you.' But he would not permit any examination, and I was very angry with him. I advised him to get his bowels well emptied every night, just before going to bed, by large common gruel or warm water injections; after that to inject twenty drops of sedative solution of opium mixed with a solution of starch. I finished by saying, 'Let me see you two or three weeks hence.' 'What,' said he, 'no pills?' 'No.' 'No medicine?' 'No.' 'Nothing at all?' 'Yes,' I said, 'do not neglect your diet; take care that the fæces shall be soft and small, and not hard or massive.' 'What, nothing but that?' 'No, nothing.

"He and his son then went away, and I was glad when they were gone. I scarcely expected to see this patient again, but towards the end of October, that is, in about six weeks, he called again to inform me that he felt quite comfortable ever since his former visit, by only doing what I had told him. Now, here is a case in point. This man's rectum had been painful for nearly three months; besides the almost constant use of figurative medicines, two or three times a week he was examined by a speculum and had injections of silver, of zinc or copper, and all that sort of thing, adding, as I believe, to the local irritation, until his condition was hardly endurable. By the simple means that I have mentioned he was at once improved. So far as I know he was cured by the method of physiological rest, as opposed to violence or physiological disturbance."

He had never seen long-continued rest produce harm; healthy joints which had been confined for months came out undamaged from months of continuously enforced rest. Yet there is one of his observations which does bear on the utility of movement in preventing anchylosis. He remarks on the rarity of anchylosis in the joints between ribs and the spine. They are joints at which movements go on unceasingly throughout life. But he refused to associate their freedom from disease and from anchylosis with their inability to assume a position of continuous rest.

If Hilton's first service to surgery was to give "rest" a foremost place in the means of treatment, his second service was to give "pain" its rightful place in the means of diagnosis. He was a teleologist; he regarded the human body as a specially designed machine provided with a sense of pain to serve as a signal of disease. "I conceive that pains situated on the surface of the body and associated with some abnormal state of an internal viscus, must be looked on as a beneficent provision, enabling us by external pain to receive the information and to appreciate slight organic changes or derangement of function of the internal viscera." That statement of Hilton's was particularly directed against those who hide their ignorance by tracing all obscure pains to rheumatism and gout. In his opinion, pain was provided by Nature to guide the surgeon to the seat of disease, and a knowledge of the distribution of nerves, and the manner in which the nerve system acted, was the means by which the surgeon was to obtain a correct diagnosis. The surgeon-detective "ran down" the disease by following up the clues provided by pain and tenderness.

Pain, in Hilton's opinion, was the signal by which Nature made known her demand for rest. He made certain observations which throw some light on the production of pain. In a case where the ulcerated articular surface of the astragalus was exposed, he made his pupils note that when he merely touched the ulcerated area the patient felt nothing, but when he pressed it there was pain; pressure gave pain; rest relieved it. In another case, one of necrosis of the tibia, he showed how pressure on the artery above the seat of disease diminished pain, while pressure on the veins increased it. Pressure within the capillaries caused increase of pain.

We have said that Hilton often construed "rest" in a wide sense. When he applied a truss to compress chronic sinuses of the groin or put pressure on abscesses of the foot, so as to bring their opposite walls into contact, he believed that the healing which followed this means of treatment was because he had given the parts rest. I shall cite an instructive instance where rest was given by removing pressure from a nerve. "Some years ago a sailor came to Guy's Hospital, having had a bad fracture of the lower part of the radius. There was no surgeon on board the ship; the fracture was not properly set, and in consequence the callus pressed severely on the median nerve above the wrist. The man had ulceration of the skin upon the thumb, the forefinger, and the middle finger. The hand was purposely flexed and put in such a position that all the tension was taken off the median nerve, and the ulceration got perfectly well; but as soon as his hand was allowed to be freely used and extended again, and the pressure of the callus permitted to exert its influence on the median nerve, the ulceration reappeared on the thumb, forefinger, and middle finger. This is a good illustration of pressure upon a nerve producing deterioration of structure." The case not only illustrates the trophic influence of nerves, but also the wide sense with which Hilton used the term "rest." To relieve a nerve from pressure or to open a tense abscess was, in his opinion, instances of the application of physiological rest as a means of treatment.

I should give a very imperfect picture of the foundation of

Hilton's methods of treatment were I to omit his inquiries into pathological conditions. I need only quote two cases to show that he sought in every case, before commencing treatment, to realise the exact nature of the lesion with which he had to deal. He never allowed an opportunity to escape which would yield him information. The first case I shall cite is that of "a young gentleman, eighteen years of age, living in Islington. On a Sunday evening, returning from chapel, wearing a very narrow high-heeled boot, walking at the edge of the pavement, his left foot turned inwards with a sudden jerk, and he exclaimed to his sister, who was with him, 'Oh, I have twisted my foot; I never had such a dreadful wrench before.' I saw him on the following Sunday with Dr. Billinghurst, and I then came to the conclusion that he had sustained some injury to the lower epiphysis of the tibia; for upon close examination, I found the ankle-joint free from mischief." Hilton advised rest. Some days later, when he was visiting patients in the City, he learned, by accident, that the young man was dead. He set out for Islington. "The hearse was at the door, so that I had only time to unscrew the coffin and examine the leg. . . . I took away the bit of tibia; it fairly rewarded me for the trouble. There was a laceration between the shaft of the tibia and its lower epiphysis, precisely the spot where the boy had the first sense of pain on the Sunday morning." Hilton thus provides us with an everlasting example of the manner in which surgery is made into a rational art.

The second case I am to cite is equally instructive. "A man going across Blackheath, rather the worse for liquor, fell down, or jumped, six or eight feet into a gravel pit, and, alighting on his feet, his leg was very severely broken. I amputated the leg below the knee. On examining the ankle-joint, which had apparently not been injured, I found the articular cartilage upon the astragalus actually depressed at one part, and at another I saw within it a large black deep patch." This patch Hilton cut out, and brought to my predecessor at this College—Mr. Quekett—who cut microscopical sections of it, and found that it was due to an extravasation of blood beneath the articular cartilage. In this instance Hilton provided us with a kind of observation of which we still stand much in need—the lesions of joints which we classify under the generic term of "sprain."

In the Museum of the Royal College of Surgeons there is a specimen which illustrates the pains taken by Hilton to clear up obscure conditions which had puzzled him while the patient was still alive. The specimen is the upper part of a spinal column laid open to show a healed fracture in the lower part of the neck. It is clear that the fifth cervical and all the vertebræ above it had slipped forwards, leaving the body of the sixth projecting into the spinal canal and pressing against the spinal cord. The vertebræ at the site of fracture have become ossified together, the projecting angularities have been rounded off, and the repair affords an excellent example of Nature's reparative handiwork. The early history of this case we owe to Mr. Whitmore of Coggeshall, a parish in Essex at no great distance from the one in which Hilton was born and bred. "It was in May 1836," writes Mr. Whitmore, "that I was called up, on a Sunday morning between four and five o'clock, to John Carter, aged 21, who had fallen from a tree, when in pursuit of young rooks. When I saw him he was perfectly insensible and motionless; cold and breathing imperfectly; with a pulse weak in the extreme . . . there was a perfect absence of muscular power and of sensibility of the skin throughout the body, except in the head and upper part of the body. The bladder was paralysed and the catheter required. There was no appearance externally to indicate the precise situation of the injury." John Carter lived for fourteen years after that Sunday morning he went in pursuit of rooks, and might have lived much longer had not the boy who "dragged him about in a little four-wheeled cart upset him so that he fell with great violence upon the ground, as he could not put out his hands to save himself." This fall set up "a chest affection from which he died in a few days." During these fourteen years John Carter "used to amuse himself or earn his living by making copies of engravings with a paint brush held in his mouth." We have specimens of Carter's artistic ability in the College Museum.

In the later years of John Carter's life, Hilton made several journeys to Coggeshall to examine his remarkable case. He found "perfect loss of sensation in the lower and upper extremities, except indistinct sensibility on the left side as far as the elbow. Muscles of the left shoulder more developed than on the right. Feels distinctly on the left shoulder and indistinctly on the right. The left forearm is now flexed; the thumb is turned into the palm of the hand and the fingers are bent over it. Right arm nearly straight; the little and fourth fingers flexed; the hands remained open until about six months after the accident, when contraction commenced. No contraction in the feet except that the right foot is a little flexed. Legs jump a little during defæcation and sometimes suddenly without obvious cause. Arms jump, especially the right, during micturition. Bowels not opened without medicine (senna). On some days has peculiar feelings of chilliness, becomes pale, and then feels hot and flushed during defæcation and micturition. The more constipated, the more these peculiar sensations are experienced. Urine very offensive when he has caught cold; at other times not so offensive, but always a little so. Urine acid. Feels a distinct pain in the bowels occasionally and now and then an aching pain in the loins. When sick, vomits with great difficulty. Erections of penis are frequent and last a quarter of an hour, with a slight escape of seminal fluid occasionally. Spine: nothing abnormal to be felt. No costal movements during respiration; no hiccough. One good meal of meat daily." Having made this graphic record of symptoms we have just given in detail, Hilton, after duly impressing the surgeon of Coggeshall with a proper sense of the importance of the case, returned to his consulting practice in London, hoping some day to learn the exact nature of the accident which had befallen John Carter and the exact manner in which its repair had been effected. He almost missed having his wish gratified, for when the patient died his friends refused to allow his surgeon at Coggeshall to examine the body, "and only upon a very special application, just before the removal of the body from the house for the purpose of immediate interment, was he permitted to take out the portion of the spine." That portion of the spine stands now on the shelves of our College Museum—a silent witness of John Hilton's love of his profession and of the indefatigable zeal with which he laboured to improve that branch of knowledge to which he devoted his life.

Hilton was a man who, if his field of vision was somewhat limited, yet saw everything which fell within that field with a keen and clear eye. No man has realised more clearly than he that the surgeon's art must be based on a knowledge of the structure and working of the human body and of the manner in which repair is effected in diseased or injured tissues. Of all the means which surgeons could command to assist the tissues to overcome disease and effect repair, he believed rest to be the most valuable. He employed his knowledge of anatomy and physiology to provide him with the means of giving rest. For him a splint was an instrument for securing rest for diseased joints and injured limbs. Hilton's long hip splint is well known. Massage, exercise, active or passive, had no place in his system of therapeutics. It is true that in cases of infantile paralysis he ordered "the whole limb to be rubbed night and morning with sperm oil and kept warm night and day by surrounding it with cotton wool or flannel." Such cases, in his opinion, were exceptional and required exceptional treatment. It must not be supposed that Hilton was the first to formulate the principles of physiological rest as a system of surgical therapeutics. In a future chapter we shall find that long before John Hilton was born, M. David of Rouen had considered the place which the surgeon must give to rest in the treatment of surgical conditions. M. David, however, gave action or movement an equally important place; there were surgical conditions in which he regarded rest as injurious and action or movement highly beneficial. In the light of another century of experience the majority of modern surgeons have come to accept the teaching of David rather than of Hilton. That is the verdict which a succeeding generation has passed on Hilton's teaching. But never again shall anyone preach the gospel of rest so freshly, so practically, or so persuasively as John Hilton.

Hugh Owen Thomas,

From a sketch made about 1884 when he was in his fiftieth year. He died on 6th January 1891 in his fifty-seventh year.

CHAPTER III

THE PRINCIPLES AND PRACTICE OF HUGH OWEN THOMAS

Hugh Owen Thomas was born when his mother was on a visit to her parents at Bodedern, Anglesey, 23rd August 1834. Her husband, Evan Thomas, was also a native of Anglesey, where his forefathers for many generations had practised as bone-setters. At the time of Hugh's birth his father had established his home in Liverpool, where his fame as a bone-setter had already spread beyond the confines of the city. Hugh was a delicate child, and, on account of his health, was brought up in the country, attending school first in Anglesey, then in St. Asaph, and afterwards, when he lived at home, Dr. Poggi's school, in New Brighton. His professional career may be divided into the three following stages :

(1) APPRENTICESHIP AND STUDENTSHIP—from 1851, when he went at the age of 17 as apprentice to his uncle, Dr. Owen Roberts of St. Asaph, until 1858, when he "returned from a short period of study" in Paris to commence the practice of Medicine in Liverpool at the age of 24. Of these seven years, four were spent with his uncle, two at the University of Edinburgh, and one at University College, London, and in Paris.

(2) A period of PRACTICE AND OBSERVATION extending from 1858,[1] when he joined his father in Liverpool, until 1875, when, at the instigation of Professor Rushton Parker, he commenced to publish his observations. The partnership with his father—a skilful, but moody man—at 72 Great Crosshall Street—lasted scarcely a year; in 1859 H. O. Thomas made his home at 32 Hardy Street; in 1870 moving to a house close by—No. 11 Nelson Street, his home until his death. In an adjacent house he established his hospital and set up his workshops. Thus his life was spent in the most central and busiest quarter of Liverpool. In 1864, at the age of 30, he made a happy marriage, but had no children.

(3) A period of PRACTICE AND PUBLICATION—extending from 1875, when he was 41, until 6th January 1891, when he died early in his fifty-seventh year. A list of his chief publications is appended, compiled according to the date at which they were issued. In publication, as in every affair of his busy life, he set convention aside and followed his own judgment or fancy. He employed his own printer and, at first, acted as his own publisher. The high-wheeled phaeton, in which he drove a pair of black horses, was of his own design; so was the nautical peaked cap he wore to shade his eyes; so were the medical theories he evolved and the surgical appliances he invented. He was a thin, dark, fragile-looking man, well under medium height—5 feet 3 inches or 5 feet 4 inches —but of indomitable spirit and of unbounded power of application. His profession was his hobby, and in his hobby he included the art of healing in its widest scope.

IT was the custom of John Hunter to prescribe "Rest" as a routine measure in the treatment of disablements of the motor system of the human body. After him came John Hilton, who regarded rest as the most powerful aid which the surgeon could bring to the aid of disordered tissues. Hilton elaborated the means of securing rest into a system; but the man whose principles and practice we are to describe here made rest his creed and ritual. Hugh Owen Thomas believed that an overdose of rest was impossible. To use the

[1] In this year the Medical Register for British Practitioners was established by Act of Parliament.

expression which he never tired of repeating—Rest must be "enforced, uninterrupted, and prolonged." I had selected the names of these three men in the above order because they appeared in British surgery at successive periods: Hunter, 1728–1793; Hilton, 1807–1878; Hugh Owen Thomas, 1834–1891; with each successive man and period rest became more urgently advocated, as the best means of assisting the natural powers of repair to overcome the violence done by disease or injury. It was only when I came to arrange the matter of this chapter that I saw that our chosen advocates of rest had been drawn into the service of surgery from three diverse racial elements of our British population. Hunter was a Lowland west-country Scot, but if his physical appearances do not belie him, he had more than a mere strain of Highland breeding in him. Hilton was a sturdy Saxon, strong, overbearing, made to rough-shoulder his way through the world. H. O. Thomas was a pure product of North Wales. He was born in Anglesey of a long ancestral line which had handed on the art of bone-setting from father to son. He spent his early boyhood in North Wales, and in 1851, at the age of 17, he was apprenticed to his uncle, Dr. Owen Roberts of St. Asaph. He had the imaginative temperament, so common amongst his countrymen, and masked a nervous and retiring disposition by a somewhat brusque manner.[1]

Men could not have approached a common objective by more diverse routes than did Hunter and Hilton on the one hand, and H. O. Thomas on the other. There was no prolonged and laborious apprenticeship in the dissecting and post-mortem rooms for him, no appointment to the staff of a great teaching hospital, with the commanding position, affluence, and social status which such appointments can and do bring to ambitious men. Thomas, we shall find, is a plain man, guard-

[1] I take this opportunity of thanking Professor Rushton Parker, Sir Robert Jones, Sir Vincent Evans, Sir J. Lynn Thomas, and Dr. Dawson Williams for their liberality in supplying me with information relating to Mr. Thomas's earlier life. For an appreciation of H. O. Thomas as an orthopædic surgeon, see articles by Dr. W. Colin MacKenzie, *British Medical Journal*, 1917, i. p. 669, and by Sir J. Lynn Thomas, K.B.E., C.B., F.R.C.S., *British Medical Journal*, 15th July 1916.

ing jealously his simple manners and homely Liverpool life, with more than a shade of contempt in his speech and deportment towards the men who sought to borrow authority from their official status and worldly position. Hunter and Hilton worked long and hard in the dissecting and post-mortem rooms in search of facts to guide them in formulating rational means of treatment; they observed the efficacy of their methods in regulated hospital wards and on closely tended private patients. The circumstances amidst which Thomas matured his experience were very different; his field of observation was the steady stream of accident cases which poured into his surgery from the dock-land of Liverpool. His field of experiment lay in his upper-workroom where, in workman's attire, and with the hand of an expert, he wrought the exact form of splint or machine which he desired for the treatment of each particular case which came under his care. Here, then, is a surgeon of a new kind, one who could and did use his knife, but it was his final and fixed opinion, founded on thirty-three years crowded with experiments on orthopædic cases, that the blacksmith's hammer, deftly used, was, in most cases, a more powerful reparative instrument than the surgeon's knife.

Hunter and Hilton never ceased gathering observations on the condition of parts as revealed in the dissecting room and post-mortem room; they kept adding to their own stock and to the world's stock of knowledge. But you will search the many writings of Hugh Owen Thomas and fail to find a single observation made by him on the state of parts seen at operation or after death; not a glimpse of the disordered state which it is the surgeon's task to bring back to a healthy condition. It is true that Thomas never did study those conditions in the post-mortem room, dissecting room, or experimental laboratory, but there never was a man who studied more persistently and observed more closely the manifestations of disease and injury as seen in the living state. For him surgery was an experimental science; each case was an experiment to be studied, diagnosed, explained, treated, and the results of such treatment to be carefully recorded. "There is an opinion prevalent," he noted, "that only gentlemen on the staff of our public charities can treat, with any chance of success, disease of the hip-joint, and certainly hitherto they have had advantages not possessed by the general practitioner." But he was also aware, as Sir James Mackenzie has persistently pointed out, that the general practitioner has opportunities from which the staff-surgeon is debarred; he sees a case in its initial stage, he takes it under his personal supervision, and can follow its progress to the end. Thomas, too, lived in the midst of one of the world's biggest industrial battlefields; the limits of his practice lay along the shores of the seven seas, for seldom a week passed without bringing him some derelict case of accident. It was Hugh Owen Thomas's great merit to have proved that a busy general practitioner can, by purely clinical methods, win for himself a permanent place among the benefactors of Medicine.

Having thus broadly defined Thomas's place amongst surgeons, it is now necessary for us to trace him through the earlier years of his professional life, and watch the development of his principles and practice. He was born and bred an orthopædic surgeon. His father, Evan Thomas, had his home and surgery, so Professor Rushton Parker has informed me, in 72 Great Crosshall Street,[1] Liverpool, where, although unqualified, he was the referee of certain workmen's clubs—dock gatesmen, ships' carpenters, boilermakers, and several others—the members having unbounded faith in his treatment of injuries of all kinds. He had five sons, all of whom became qualified surgeons; Hugh Owen was the eldest.[2] As the boys grew up they had to help in the surgery in Great Crosshall Street, and could not fail to gain a first-hand knowledge of injury and disease. Indeed, it was proposed at one time that Hugh should become his father's apprentice, but the ill state of his health—he never at any period of his life was robust—led to a change of plan, and in 1851, at the age of 17,

[1] Later, he had his home at Seacombe on the south bank of the Mersey, but retained his surgery in the small house in Great Crosshall Street.

[2] Mr. Evan Thomas still lives and practises in his father's old home, 72 Great Crosshall Street. He was born in 1838—four years after his celebrated brother Hugh.—RUSHTON PARKER.

he was apprenticed to his maternal uncle, Dr. Owen Roberts of St. Asaph, 25 miles from Liverpool as the crow flies. His apprenticeship lasted for four years. In 1855, at the age of 21, he went to Edinburgh to study Medicine. Syme and Spence and Simpson were in the heyday of their fame; Goodsir was teaching anatomy, there was then a demonstrator of anatomy newly arrived from London—William Turner. Lister was also there commencing the Edinburgh phase of his great career. But in Thomas's writings there is never an allusion to his life in Edinburgh; only one man seems to have impressed him, Hughes Bennett. He apparently spent two winter sessions in Edinburgh; in 1857, his third session, he attended classes at University College, London; of his life and experience in London he never gives us the slightest hint in his writings. In 1857, in his twenty-third year, he became a member of the Royal College of Surgeons of England, and then proceeded to watch the practice of French surgeons in the hospitals of Paris. In the spring of 1858, the year which saw the Medical Register established in Great Britain and Ireland, he returned to Liverpool to help his father in the Great Crosshall Street surgery. Of what he saw and learned in that surgery during his boyhood, in the vacation periods of his apprenticeship and student days, and now during the short year he was to spend in his father's practice as a qualified practitioner—for before twelve months were finished, father and son separated—he gives us several interesting and instructive glimpses:

"Many cases have I observed enter the consultant's surgery, lame and in pain, who after being well fitted with a layer of stiff adhesive plaster, over and around the affected articulation, left the surgery less lame, and in less pain; sometimes even without pain. This is the history of some; others indeed had a different termination."—*Review of Past and Present Treatment*, 1878.

"For many years I was a witness of the treatment of joint disease by methods sometimes purely expectant, at other times consisting of a fractional fixation, and the results in some instances were certainly so striking as to excite my envy. But these very cases I now know would have recovered, some with no attention and others with but imperfect rest. For one result which excited my admiration ten failed. What is wanted is a method which will benefit all cases."—*Review of Past and Present Treatment*, 1878.

"I watched for many years the extensive practice of an untrained gentleman and, to do him justice, record here that he never failed to secure a perfect restoration of these fractures (of lower end of radius), and if I were to say that I observed his treatment of 200 cases, I should be vastly underestimating their number; yet his treatment, guileless of anatomy, was only thorough reduction and reasonably applied counter-pressure. . . . He was equally successful in his management of fractures of the lower end of the tibia up to the period of use, when, from his not taking precaution to centralise the weight of the body on the foot, several of his cases would lapse into deformity."—*Contributions to Surgery and Medicine*, 1887, Part III.

These extracts show us that when he left his father and set up for himself in 32 Hardy Street in 1859 he had already acquired an extensive and very practical knowledge of a difficult class of surgical cases—a knowledge which was being checked and criticised under the light of the medical experience gained in Edinburgh, London, and Paris. It is but a short distance from Great Crosshall Street to Hardy Street, and most of the club appointments, worth about £500 per annum, which had been held by his father, followed him to his new home. He had no waiting period; he started away with an assured practice. We have to watch him commence his life's work there if we are to understand what he succeeded in accomplishing in the thirty-two busy years which are to follow. He is up early in the morning—at 5.30; before breakfast at eight he has already made a round among his patients; he is in his surgery from nine to one; he has an afternoon round, an evening spell in his surgery, and again another late round of visits. He follows the progress of his cases very closely; they are his hobby. He already has a theory or principle which he is applying with all the might and ingenuity of which he is capable. He rarely mentioned Hunter in his writings, but the principle he is to apply—even the language in which he couches that principle—is truly

Hunterian. The following extract is his statement of the principle which guided him:

"If any person, with a part diseased, possesses sufficient vitality so that there be a tendency to reparation in the diseased locality, then Nature always has a mode of operation, and in very many instances the natural method of restoration has become known to us. Indeed the practice of medicine and surgery mainly consists of either aiding or controlling or supplementing this natural effort to resolution."—*Contributions to Surgery and Medicine*, 1883, Part II. p. 7.

It is clear then, when Thomas set out from 32 Hardy Street to sail the sea of practice, he already realised that he could not cause the tides or raise the winds that brought derelict human ships to the harbour of resolution; he realised that the utmost he could do was to steer in such a way that the best use was made of such tides and winds as came along. Even at the end of thirty-two years of incessant observation, as the following extract will show, he was often in doubt as to how far a happy issue was owing to his seamanship, or how far it depended on the fortune of the elements:

"As in Medicine, so in Surgery, even a discerning practitioner often finds it difficult to satisfy himself whether the patient got well with his assistance or despite his interference" (1890).

In Hardy Street he lived for eleven years; there he married and prospered in spite of ill-health, overwork, and overstudy, and in 1870, at the age of 34, moved to a house close by—11 Nelson Street—which his nephew, Sir Robert Jones, still occupies. There he held his Sunday clinics. "When the bells were tolling for church, the surgery at Nelson Street was filling with a congregation of suffering folk." Thus he lived from his twenty-fourth to his fortieth year, from 1858 until 1874, his fame during that period being still confined to dockland. In 1874 an event occurred which led to his labours becoming better known. Mr. Parker, a police surgeon, had reason to visit Mr. Thomas, who at that time had his surgery and hospital in Nelson Street, in connection with a case—that of an inspector of police—who was then under Mr. Thomas's care, on account of a compound fracture of the leg. Mr. Parker had taken with him his son, Rushton, a young surgeon of 28, with an assured future in the Liverpool School of Medicine. We can learn the effect of that visit from the preface which Thomas wrote for his first important publication on *Diseases of the Hip, Knee, and Ankle Joints, with their Deformities: Treated by a New and Efficient Method.* "I saw at once," writes Professor Rushton Parker, "that here was a master, and the acquaintance ripened quickly into friendship, I insisting on his publishing immediately an account of his hip and knee splints, because, I said, 'these are things I must use, teach, and publish; and there are men in this town who will palm them off as their own if they get to use them intelligently.'"

That was Thomas's first important publication, but not actually the first. In 1867 he had published a short paper in the *Lancet* on the treatment of un-united fractures, and in 1873 described a method of using silver ligatures for the treatment of compound fractures of the mandible. The ligatures were fixed by a terminal spiral which could be conveniently tightened as the mandibular fragments yielded and came together. He also advocated then, the practice adopted in the present war, of removing any tooth situated on the line of fracture, as its presence was always a cause of retarded union. Once started as a writer, he continued to issue *Contributions to Surgery and Medicine* almost annually, until death brought an end to a career handicapped by indifferent health and overtaxed with work. He died in 1891 at the age of 57. Even in the manner of publication he emphasised his individuality. His writings were issued in pamphlet form, designed, one would almost think, for a temporary and private circulation; but time will show that they have a permanent value for thoughtful medical men. It was his habit to reprint, at the end of each contribution, the press notices and strictures which had greeted its first appearance. On very few occasions did the critics recognise their merits or extend a welcome to them. It is because of the means he adopted for publication that his numerous writings are so little known. It is very difficult now to obtain a complete set of Hugh Owen Thomas's Works.[1]

He based his methods of treatment on the axiomatic belief which had served to guide Hunter and Hilton: viz. that the power of repair is an inherent property of living tissues. "The crying evil of our art in these times (1883)," he writes, "is the fact that much of our surgery is too mechanical, our medical practice too chemical, and there is a hankering to interfere, which thwarts the inherent tendency to recovery possessed by all persons not actually dying." Further on, in another passage, he adds: "There are actions which Nature cannot do so well as the artist in charge." He held that in only one way could the surgeon aid the reparative powers of injured or diseased tissues to come into full operation, and that was by giving the part rest. His message to the surgeons of his time was that they did not understand the meaning of the word "rest." Hilton, he said, fixed a limb in a splint, and believed he had given it rest. Immobilisation he held to be the first requisite, but it must be applied in such a way that the diseased part was not compressed, nor the normal circulation of the limb in any way interfered with. All forms of plaster-splint necessarily exert an injurious pressure, and for that reason and several others he abandoned them at an early stage of his practice. Pressure, he held, was a form of restlessness or irritation. If he applied a stiff or elastic bandage to a diseased knee-joint, so as to compress and surround it, he held that in so far as he lessened the movement of the joint he assisted repair; but in so far as he compressed it he hindered repair. He distinguished such a form of immobilisation as that of "direct fixation"; his ideal form of immobilisation was that of "indirect fixation," which is best illustrated by his knee splint. He designed this splint so that it could prevent

[1] "The printer, a quaint character, whose name was Dobb, lived in a small shop in Gill Street. He was factotum and publisher. In the later editions the name of H. K. Lewis appears on the covers. Very few were sold, and the remainder occupied a large room in Mr. Thomas's house in Nelson Street."—Sir ROBERT JONES, 1913. See list of publications at end of this chapter.

movement at the knee, and yet leave the joint uncompressed and the circulation of the limb unhindered. All forms of immobilisation which depended on traction by weights, pulleys, or elasticity he discarded, not only because they transgressed his conception of rest, but also because in practice he found them less effective than that of "indirect fixation." Rest had to be continued without interruption until all trace of unsoundness had disappeared from the joint, and then, that point being reached, the cure would be completed by the gradual return of natural voluntary movements. Therein lies the essential doctrine Thomas came to preach to the surgeons of his time.

If I were to cite an instance to illustrate how his conception of rest and the manner in which his methods of treatment differed from those of all the men who had gone before him, I should select it, not from his papers on fractures and diseases of joints, but from one which he regarded as of great and of lasting merit, that in which he describes the best means of treating cases of obstruction of the bowel. In his time, such cases were usually treated by active measures, by purges, enemata, kneading, and movements; as a last resort enterostomy was performed. He had followed cases of acute obstruction of the bowel during his apprentice days at St. Asaph, sometimes to a successful at other times to a fatal termination, and at a later date began to apply his treatment of rest, with a success which led him, in 1875, to write his first paper on this subject. It was not, however, until 1884, when the treatment of such conditions became a subject of discussion at the Liverpool Medical Institution, and afterwards of an acrimonious correspondence in which Sir Mitchell Banks was involved, that this part of his labours received public attention. In his treatment of such cases we obtain a concrete illustration of what he means by "rest." The treatment of the case I have selected for an example has been placed on record by his nephew, Sir Robert Jones.[1]

"I remember well how Mr. Thomas called the relatives

[1] *Intestinal Obstruction Thirty Years Ago.* Presidential Address, Liverpool Medical Institution, October 1913.

together and told them he was going to make a fight for the patient's life. He urged them to be loyal and to help him, and, to add to his persuasion, he threatened them with a coroner's inquest if *they gave the patient anything* without express permission. Above all things the patient was to have no milk; that curdled and loaded the bowels with solids. Then the foot of the bed was to be elevated to lessen the pressure in the abdomen; a morphia injection was given, and nothing but sips of water with a little arrowroot. For the first few days he visited the patient five or six times a day. Almost immediately the vomiting was reduced to about once in twenty-four hours, the patient became easy and slept, but the abdomen was tense. Twice he performed paracentesis (for the relief of pressure), being careful not to allow the trocar to remain in the intestine longer than a few minutes, for fear of a fistula. On the twenty-fourth day, at intervals, a very little flatus was passed, on the twenty-sixth large quantities, and later in the day a few small scybalæ. Again, during the night a copious pultaceous motion, and this again followed for three days by prodigious quantities of thin fæcal fluid."

From the method in which that case was treated we see that rest was to be applied with the meticulous care and rigidity of a Calvinistic doctrine. Rest was to be secured first, by sedatives—to place the bowel at ease and relieve pain; the bowel was to be restrained from all manner of work by absolute starvation; he declared he had never seen starvation cause death in a case of intestinal obstruction, however prolonged. The bowel was not to be disturbed by any act whatsoever, such as the giving of enemata or "rectum tickling," as he most unprofessionally phrased the practice. "Nature," he said, "is late in working a relief, and patience is needed." His critics said his treatment of intestinal obstruction was not new, and modern surgeons will declare it to be bad. His medical critics, however, were wrong; opium and starvation had been often prescribed and employed in such cases; rest had been enjoined, but Thomas was the first to apply rigidly and completely the principle of rest as a logical system to such conditions, and to carry it out in the form of "enforced, uninterrupted, and prolonged rest." In cases of fracture of the femur, in order that the patient might escape the disturbance caused by the act of defæcation, he placed the bowel at rest by sedatives and low diet, but never succeeded in keeping it still beyond the twenty-first day. There is all the difference in the world between the pious enunciation of a principle and its strict, systematic, and complete application. Therein lies the difference between Thomas and his predecessors.

It has been said that Thomas's methods of treatment are not founded on a knowledge of anatomy, physiology, and pathology—the only basis for a rational means of treatment. That, in one sense, is true. "Surgery is an experimental science," he states. All his life long he kept making experiments in his methods of treatment, closely observing and carefully recording the results night by night. The conclusion he drew from these years of toil was that the more strictly he observed the principle of rest in his method of treatment, the better were his results. But it would be a complete misrepresentation of the case were it to be said that his methods were not based on a knowledge of the anatomy and physiology of the human body. "Men admired my splints as if I were a blacksmith," he wrote, "but the principles on which they were framed they never could see." We shall examine those principles in connection with the hip-joint. We shall find he is to reintroduce to us a very old kind of anatomy, physiology, and pathology. Thomas never thought in terms of muscles, but of parts; when in ordinary-day life we execute movements, however perfect our knowledge of anatomy may be, we do not think of the muscles which produce them, but only of the segments or parts of the limb or body we wish to move. The anatomist or physiologist who would sing, play golf, or acquire any skilled series of movements is well aware that his professional knowledge gives him no advantage; the teacher who corrects his faults knows nothing of muscles—only of right and wrong movements. Yet that teacher is a real anatomist and physiologist, and it is in that sense that Hugh Owen Thomas was an anatomist and physiologist of a very rare kind. When he came to apply his doctrine of rest to the hip-joint, he saw that it could not be done by applying the long splint to the side of the body and lower limb—even if it stretched from the arm-pit to the external malleolus. Such a splint lay outside the axis of the hip-joint; it was separated from that axis by the lever represented by the head, neck, and great trochanter of the femur. It could not be fixed to the side of the chest because there the splint crossed the ribs at their point of greatest movement. To cross the axis of the hip-joint, so as to secure a complete control of its movements, the splint must be applied from behind and carried up the back where the ribs are at rest, for the dorsal segments of the ribs undergo only a rotatory movement where they are covered by the spinal musculature. He therefore applied his hip splint to the dorsal surface of the body and thigh. He realised that no two people are shaped alike and that the splint must be accurately moulded for each patient, and only one with a knowledge of anatomy could apply it. He therefore chose a pliable material—wrought iron—of sufficient strength, and invented the tools by which it could be shaped, and adjusted the splint to the patient's body with his own hands. He considered that the fitting of a splint was the surgeon's duty.

He observed that in all movements of the lower part of the body—in walking, turning, or bending—two joints were correlated in their action, the hip and lumbo-sacral joints of the spine. He saw that a limitation in the movement of the hip could be compensated for, by an increased action in the lower part of the spine, and on that observation he founded his test for demonstrating and estimating the degree of limitation in the movements of the hip-joint. He realised that the hip-joint could not be fixed, could not be given rest, unless the dorso-lumbar region of the spine were also fixed. Movements of the knee also affected the hip; the knee had also to be fixed to give the hip rest. To fix these joints he invented his hip splint—and fashioned it in such a way that it could be fitted and attached to sound parts of the body and could be so placed that it would assist the weaker groups of hip-joint muscles against their stronger antagonists. He therefore placed the main supporting bar of his hip splint along the back and prolonged it to fit the flexor aspect of the thigh and leg, bent so as to support these parts and relieve the flexor muscles of the hip from those burdens which disease had thrust upon them. He did not know why, but he did know that these muscles would begin to relax so soon as they found they were relieved of their involuntary burden. But he did much more: he saw that the splint must be perfect in its fit, and its effectiveness; he attended to that day after day and week after week, moulding the splint as the muscles relaxed, thus leading the parts down the ladder in ease which they had slowly climbed in pain. We see that he carried out, just as he proposed it should be carried out in the treatment of intestinal obstruction, and as it never had been carried out before in cases of diseased hip-joint, a system of "enforced, uninterrupted, and prolonged rest." "A man," he said, "who understands my principles, will do better with a bandage and broomstick than another can do with an instrument-maker's arsenal."

His discoveries and skill as an applied physiologist we can illustrate also in connection with the hip-joint. A hip-joint which possessed the normal range or radius of voluntary movement was in full health; its articular surfaces, ligaments, bone, muscles, and nerves could not then be the site of injury or of disease. If the movements were limited, then some part of the machinery of that joint was affected by injury or disease; the degree of the limitation was a sure index of the extent of the mischief. If it was found that on comparing the limits from one week to another there was a greater degree of restriction, then that joint was undergoing an increasing degree of pathological change; if, on the other hand, the voluntary movements were less restricted, then that was a sure index that resolution was at work. He invented accurate tests for estimating the extent of movement at the hip-joint by placing his patient in a supine position on a flat surface so that he could observe and eliminate the lumbo-sacral movements before estimating the mobility of the hip-joints. Pain, heat, swelling, and redness had their value as diagnostic guides, but for him the most delicate and practical test for disease or derangement was

4

that of use—of range of movement. Over and over again he was able, by the application of these physiological tests, to recognise incipient stages and latent residues which escaped the eye trained in orthodox methods. He was indifferent as to how far that limitation was due to reflex effects, or how far they depended on voluntary protective efforts on the part of the patient; in either case the muscles had to be put to rest, and for that there was but one means—the hip splint he had elaborated on a definite anatomical and physiological basis. One of his physiological tests he rightly regarded as of the greatest value—the one he applied when he removed a splint from an injured joint or limb. If the patient after removal of the splint had a power of movement and could by an effort of will poise the limb exactly in the position in which it had been fixed in the splint, that was evidence of complete muscular control and of sound health in the part.

In 1883, knowing nothing of Wolf's law, he enunciates the doctrine that "Time and Physiological Action will commode the part to the direction of the employed force" (*Contributions to Medicine*, Part II. p. 121). From beginning to end we see that his orthopædic practice is founded on the truth of that law. Of greater importance, from a practical point of view, is his doctrine of "unsoundness." Much of his practice is founded on this doctrine. An unsound part is one which is the site of disease or injury; it is one in which inflammatory processes are taking place. For the orthopædic surgeon a state of "unsoundness" is of the greatest import, because such a part is then plastic and can be moulded. The deformed knee was most easily straightened when it was in an inflammatory state, and the badly set fracture when it was in a stage of healing. We see him apply this doctrine to the rectification of deformities of all kinds. The deformed part—be it knock-knee, an ununited or a mal-united fracture or a deformed but "sound" joint—has first to be reduced to a state of unsoundness before it can be rectified. He had many methods of setting up a physiological state of unsoundness in a part —by percussing it, wrenching it, or applying the process which he called "damming."[1] His application of such means for the treatment of deformities was the corollary of his law of "Rest." Rest placed tissues in the most suitable circumstances for reparative processes becoming effective, so that the part became sound and fit for use. If the opposite condition was desired, then unrest—irritation—would render the part unsound and plastic; the diseased tissues would be forced to make unnatural efforts to effect repair. Movement in a joint caused an exuberant effort that ended in anchylosis; movement or irritation at the site of fracture resulted in an exuberant callus. If, then, he wished to reduce a deformity, he excited unsoundness in the area in which the deformity existed. For a like reason, if he wished to excite a sluggish tissue to action as in an un-united fracture, or one in which union was delayed, he irritated the tissues by various physiological methods—chief of which was percussion at repeated intervals by a masked hammer or the application of constriction above and below the site of the fracture or lesion ("damming"). He believed by thus hampering the circulation of a part he irritated it, set up in it a condition of tissue unrest and evoked increased efforts towards repair. He applied the same form of treatment to recurrent dislocations of the shoulder-joint, and believed that a similar treatment was suitable at the knee-joint if there was a tendency for the internal cartilages to become dislocated.

When he came to design his knee splint he utilised his intimate knowledge of the living human body. He was "guided by his memory of the natural outline of the part." Nature has two methods of providing limbs with the degree of rigidity necessary for movement: in vertebrate animals she utilises an internal or central support; in crustacea and insects a peripheral or ensheathing support answers her purpose. We

[1] Thomas introduced the practice of damming for the treatment of cases of delayed un-united fractures in 1876. In 1903 Bier introduced this method which Thomas had constantly taught and used as a means of treatment from 1876 until 1891—but British and American surgeons speak of it as Bier's method!

have no reason to suppose that Thomas borrowed any suggestion from the crab or lobster, yet his knee splint is based on the ensheathing principle, exemplified in their limbs. He simplified the sheath design by cutting away all unnecessary parts, leaving only a basal or inguinal ring, with inner and outer bars to represent the complete sheath. He saw that he had to utilise the natural base of the limb—the prominent and fixed parts of the hip bone—as a base for the new skeleton with which he was to supply support to the damaged extremity. He therefore left intact the basal ring of the sheath, and by an exact study of the living parts moulded the inguinal hoop to the supporting points of the pelvis—the ischial tuberosity and projecting iliac parts of the pelvic basin. He gave his sheath rigidity at its distal or pedal end by uniting the lateral supports. He fixed the foot in the distal end so as to keep the limb fully extended and incapable of movement. He thereby left the circulation of the limb free, and such pressure as was necessary for fixation or for reduction of deformity was applied to healthy parts. He thus furnished the lower limb with a new and temporary skeleton which relieved all its bones of work and stress, and gave the muscular engines complete rest.

It is when we watch Thomas undertaking the treatment of deformities of the foot that we recognise in him a true, if unconscious, disciple of Hunter. He bases his treatment on a knowledge of function, and with the sure touch of genius hits at once on the simplest and most effective means of putting his principle into practice. A woman, Mrs. E., brings her son for treatment in an early stage of flat-foot. He informs the mother first, that "steels" are not required (she had thought otherwise); secondly, that the remedy for the existing condition is *to assist the boy to continue the inturn of his feet*; thirdly, that hitherto the boy was making, by his will, directed to his calf muscles, an effort to avoid deformity; fourthly, that these muscles, subjected to too continuous an effort, would become tired, then probably the foot might rapidly splay, other muscles then coming into play, but only to aggravate the difficulty: that the remedy is simply to provide a particular form of boot heels, and that *leather will do the labour* which, at present, is continually thrown on certain muscles, which are not only doing their own work, but are sustaining a weight which ligaments[1] for a time are unable to sustain. The mother, however, remained unconvinced. "What!" said she, "only to put a bit of leather on his shoes and still make him walk with his toes turned in?" Thomas made another attempt "to explain why merely wearing a moderately high heel well sloped, so that the outer depth of the heel of the boot should be three-quarters of an inch and the inner edge one inch, was all-sufficient for a pair of feet in the initial stage of splay." Thomas learned "crooked-heel" method of treatment from watching the development of flat-foot in patients who had sustained Pott's fracture. In such cases he assisted the overstrained muscles of the instep by "crooking" the heel of the boot, so that the weight of the body was thrown towards the outer side of the foot. But the woman, like Naaman of old, "was wroth and went away." In an equally simple and effective manner he gave rest to the metatarso-phalangeal joints of the foot, particularly in cases of hallux valgus, by the simple means of throwing a raised bar across the sole of the boot. By this simple method the strain in walking falls directly upon the heads of the metatarsal bones instead of upon the toes and their basal joints.

The instances just cited to illustrate Thomas's "principles and practice" have been chosen from the lower limb. As regards the upper limb, we see him again bringing his practical knowledge of anatomy to bear on the shoulder-joint. As at the hip, he realised that in all shoulder movements two joints were involved: the scapulo-thoracic and the scapulo-humeral. It was his recognition of this fact which led him to plan a new means for the treatment of dislocations of the scapulo-humeral or shoulder-joint. That joint could not be manipulated with precision until the scapula was fixed.

[1] In the normal foot muscles alone maintain and balance the arch. It is not until the muscles give way that a load is thrown on the ligament. Ligaments are incapable of sustaining a steady weight without undergoing elongation.

He again utilised a basal ring—built into the side of his dislocation chair—against which the scapula could be fixed when the arm was drawn through the ring for the purpose of reducing a dislocation of the shoulder-joint. Further, we see he has grasped, in a way no one had done before him, the essential differences between lower and upper limbs. The lower limbs are framed and constituted to serve in the support and locomotion of the body; their muscles are fixed and tuned to serve these functions. Hence when he sought to rest their muscles, he fixed the limb in long rigid splints—for preference his knee splint. But in the upper limbs the anatomical and functional conditions are totally different; the arms are framed not for support but for free movement; they have no rigid basal girdle to which a fixation apparatus can be applied. As usual he fell back on the simplest means—his "gauge-halter"—which in the upper limb takes the place of the knee splint in the lower limb. His gauge-halter is simply a neckerchief tied round the neck so that its loose ends fall down the breast. The loose ends are tied so as to form a sling for the support of the flexed arm, but tied to the wrist and sealed so that neither the patient nor his friends can take liberties with the surgeon's treatment. "This gauge-halter," he writes, "when employed by me is never varied or undone until the patient is cured. Rest has to be enforced, uninterrupted, and prolonged. It is a very simple surgical appliance, but one which I frequently employ to the exclusion of all other mechanical aids. For fractures of the condyloid portion of the humerus it is, after due consideration and observation, my firm conviction that slinging the arm by a gauge-halter only is the same treatment. I am very wary that none but myself meddle with it, and examine periodically the arrangement of the cord part, especially if the case is not doing well. A wily female deceived me on one occasion by cutting the gauge behind the patient's neck and stitching it up again. After my detection of her, I always examined the cord behind the neck and have since caught several offenders."

We thus see that although he completely alters the means

he adopts in the treatment of injuries and diseases of the upper limb, the principle remains the same. He gives the part rest and secures a free circulation of the limb. His change of means was determined by the essential differences in structure and function which exist between the upper and lower limbs —differences which many of his followers have failed to realise.

To form a just appreciation of the writings of this Liverpool surgeon, it is necessary to know something of the various movements which swept across the practice of surgery during the period covered by his professional career. When he became an apprentice to his uncle in St. Asaph in 1851, the use of anæsthetics was becoming established, but the more his experience became extended the less did he rely on the inventions of Morton or Simpson. He had been practising for seven years in Hardy Street when Lister, in 1866, announced his discovery of wound-infection, a discovery which ultimately revolutionised operative surgery. That discovery scarcely affected Thomas's practice. "For some years previous to the introduction of the antiseptic method, I practised the open method and was well satisfied with the results obtained, but on the publication by Professor Lister of his successes, I at once commenced the practice of antiseptic surgery and continued to practise it for three years, with the result of being perfectly satisfied that its merits have not been over-stated nor the trouble necessary for carrying out the details exaggerated. I returned, however, at the end of that time to the open method, and have since laboured to improve it, so much so that I am emboldened to assert that the open method, in results and successes, is equal, if not superior, to anything to which antiseptic treatment has yet attained." In cases of compound fractures he laid open all the recesses of the wound and washed them out with water, or salt and water. Thus, of the Listerian movement and all the antiseptic observations which it entailed, Thomas was little more than a spectator.

Another movement which was initiated in New York by H. G. Davis, C. F. Taylor, and Lewis Sayre during the early sixties brought out certain critical and pugnacious qualities of which Nature had given him a liberal endowment. He was convinced he had grasped the great principle of rest more completely than they, and had evolved much more effective means of applying that principle in the treatment of injuries and diseases of bones and joints. Yet he saw his British colleagues extol their methods and appliances and pass his own efforts by with neglect. The New York school were advocates of extension as a means for fixing and resting disordered limbs. Thomas said the practice had been tried in England for thirty years, and was abandoned because it was wrong in principle, and the results obtained in practice by its use were bad. His language and epithets were certainly not calculated to further the objects he had in view. "Several of my friends," he wrote, "have expressed their objection to the habit I have indulged in, of criticising the treatment of others. My answer is that the errors of past practice must be laid bare, otherwise the reformed treatment is apt to be leavened with the errors of the old" (*Contributions*, Part III. p. 104).

Another movement which occupied the close attention of surgeons in Thomas's time was the excision of joints. He knew very well that, fifty years before he was born, this conservative form of surgery was conceived and put into practice by Henry Park in Liverpool Infirmary. As a student in Edinburgh he heard Syme extol its benefits; when he was in full practice as a surgeon, James Spence was proclaiming that its thorough application would open up a new world in surgery. In Thomas's opinion, the world which excision opened up, was one of unnecessarily maimed limbs—limbs which could have been saved, or at least cured equally well, had the surgeon been content to play the patient part of Nature's assistant. How he came by that opinion the following extracts will show:

"But as I dwell in a large town, endowed with several large hospitals, in charge of enterprising surgeons, who, inspired by the spirit of the profession of our times, prefer to cut mechanically what could be unloosed physiologically, I have thus been enabled to notice what can be gained by excision of diseased articulations."[1]

"Excision was very successful, so far as not being attended with a high mortality, and the limb after excision was as superior to an artificial limb as a limb cured in useful form, though with a defective joint, is superior to a limb in which a joint has been excised."[2]

In brief, he believed, as a result of prolonged observation, that there was no place for the operation of excision of joints in surgery; the limbs which could be saved by excision could also be saved, but in a much more useful state, by the application of rest, uninterrupted, enforced, and prolonged. If rest could not save them there was no alternative short of amputation. In estimating the merits or demerits of excision, he pointed out more emphatically than anyone had done before that there was a natural history of joints. The elbow-joint had a recuperative power beyond all other joints of the body, yet even in that joint, good as the results of excision might often prove to be, he held that the cures obtained by "natural means" were superior in every case.

It was during Thomas's time that manipulative surgery or "bone-setting" had one of its periodical phases of elation. In 1867, Sir James Paget, impressed by the writings of Dr. Wharton Hood and by his own observations, commended, with the weight of his great authority, the methods employed by bone-setters as the most effective means of treating limbs and joints which had become stiff from injury or from disease. Now there was no one in England so well qualified to express a reasoned judgment on the efficacy of methods practised by bone-setters as Hugh Owen Thomas. In this matter it is best to let him speak for himself.

"For many years after the commencement of my experience in surgery, I had the opportunity of observing the practice of those who had acquired a good reputation for skill as successful manipulators. Their forcible operations and

[1] *Contributions*, 1886, Part VI. p. 93.
[2] *Ibid.*, 1883, Part II. p. 147.

passive motions were supposed either to lead to, or hasten on, the recovery of joints injured or otherwise unsound. I have resorted to these performances, and for many years believed that my interference assisted recovery. Long ago I have, from a more complete knowledge, confirmed by crucial tests, so selected them that I cannot find suitable cases upon which I would perform the deception known as passive motion. And, whereas in the earlier days of my experience I believed that much aid was given to recovery by passive motion, now I know, by well-attested facts, that some of my marvels of my past practice had been marred by the very treatment I was so proud of."[1] That was the conclusion he reached after a life-time of close observation not only of cases which had been under his own care and under that of his father, but also cases treated by bone-setters of great repute.

His abandonment of "manipulative surgery" and of passive movements for the recovery of disabled joints was not altogether a result of his experience; his experience was supported by every observation he could make on the state of diseased or injured tissues. He could not see how it was possible for movement to do otherwise than increase the irritation of the part, to aggravate the inflammatory process and to increase the amount of exudate thrown out—the very substance out of which adhesions are formed. He knew from direct observation that such was the case. Nor could he conceive how the disease or injury could be lessened by passive movements, massage, or friction; no one has ever shown how they could be. But there is proof that rest favours the reactive and reparative powers of tissues and lessens the formation of adhesive matter. The principles which guided him in practice chimed with his experience; surgeons were misled in favour of manipulation by three circumstances: (1) a lack of observation on a sufficient number of control or passively treated cases; (2) a lack of patience; (3) the lack of a test to tell them when the joint was sound. He knew that once the part was sound the muscles would again come into action; they could not act normally until the diseased

[1] *Contributions*, Part VI. p. 66.

condition had been replaced by one of health. When once that state was reached he had no fear of adhesions; normal volitional movements would break them down more surely than the most skilled manipulations. The following quotation illustrates the forcible way in which he expressed himself on this matter: "It would indeed be as reasonable to attempt to cure a fever patient by kicking him out of bed, as to benefit joint disease by a wriggling at the articulation."

Like John Hunter, his opinions and practice altered as his experience and knowledge increased. "My present opinion," he wrote, "concerning late incisions for pus within the elbow-joint is contrary to the one which I held fifteen years ago; but my change of view was gradual and unavoidable, being the effect of clinical observation." We have seen that he tried manipulative surgery and abandoned it; he did the same as regards the Listerian method of treating wounds; he abandoned aspiration of fluid or blood from joint cavities after years of trial. Perhaps the best example of his change of practice relates to his treatment of fractures of the neck of the femur, of the patella, and of the olecranon. From being the lesions he was most afraid of, they became the ones which troubled him least. It is not necessary here to describe the appliances he used; his aim was to secure (1) perfect fixation of the part to ensure absence of friction; (2) freedom of blood supply; (3) an attitude of indifference as regards obtaining perfect symmetry. We see the critical and sceptical temperament with which he was endowed from the following extract: "Indeed, my experience incensed me with doubts, whether surgical treatment (in fracture of the patella) was hitherto any aid to repair, and with a belief that it was only of some help in bringing about restoration of symmetry. Observing the results of the practice of others and my own, I was forced to the conclusion that perfect restoration of symmetry was of no value towards gaining a rapid cure and a useful result. . . . It was thus becoming evident to me that the acme of treatment would be the adoption of some means that would be purely supplemental to the natural or spontaneous mode of recovery, so that the inherent tendency to repair would not be hindered" (1879).[1] Hence in fracture of the neck of the femur he applied his hip splint, with rest in bed; in fracture of the patella, his knee splint in the form of a walking caliper; and in simple fracture of the olecranon he slung the arm at an obtuse angle in his "gauge-halter." But as regards treatment of muscles his practice never changed. He speaks of them as John Hunter did, as if possessed of a species of intelligence. The "intelligent muscles, finding their labours no longer needed (after application of his hip splint), take a rest until invited to enter again on duty." . . . "If a muscle manifests a flicker of movement under volition, it will recover its full power," is another instance of physiological observation. He treated muscles as he treated his patients. He relieved the weak and oppressed; he restrained the strong. If the flexors of the elbow were paralysed, he saved them from their overstrong antagonists, by keeping the elbow partially flexed; he dorsi-flexed the wrist-joint by a special splint to nurse the paralysed extensor muscles of the forearm.

In truth, Owen Thomas was the lineal descendant of John Hunter and had the misfortune to appear in a period when the Hunterian traditions were overshadowed by brilliant and great advances made in many departments of medical knowledge. The microscope seemed to reveal a world of disease for which Hunterian methods were out of date, whereas it was the same world of disease looked at a little more closely. It was a period in which the surgeon boldly essayed to play an active, not a passive, part in healing, and in such a company there was no place for Thomas. It is true that he had not studied living matter as Hunter had, but that very inventive genius which permitted Hunter to plumb some of Nature's deepest secrets was also given to Thomas, who spent it in the design of appliances to secure the principle dearest to his heart—physiological rest. If Thomas had spent that gift on commercial projects he would have been one of the most prolific and successful inventors of his time. Yet were I to emphasise the greatest legacy he has left to Medicine, it would not be his splints or appliances, his principles, his practical applications

[1] *Contributions*, Part VI. p. 57.

of anatomy or of physiology I would underline, but his personal care in the service of his patients. No trouble was too great for him if his attention was needed to effect a cure. I will cite only one example—one which illustrates not only his peculiar ability as a biological surgeon, but his remarkable gift of taking pains:

"A young girl consulted me who was much annoyed by the slipping of her patellæ over the outer condyles whenever the knees were bent. It appeared to me that, if I could enlarge the upper surface of the outer condyles, then the knee-caps could not ride over them. After the right and left outer condyles had been moderately percussed *every week during five months*, the left patella ceased to ride over the outer condyle, and after nine months' percussion, the right patella also ceased" (*Contributions*, Part VII. p. 53).

If genius is a "capacity for taking pains," then Thomas was a genius; no one who will take the trouble to ascertain what he did, and the circumstances under which he accomplished his life's work, will fail to see that he has earned himself a place among great British surgeons.

List of Hugh Owen Thomas's Chief Publications

1875. *Diseases of the Hip, Knee, and Ankle Joints, with their Deformities: Treated by a New and Efficient Method.* First Edition, 1875; Second Edition (with Introduction by Professor Rushton Parker), Liverpool, 1876, 283 pp. 26 plates; Third Edition, 1878.

1878. *A Review of the Past and Present Treatment of Disease in the Hip, Knee, and Ankle Joints*, 1878, 66 pp.

1881. "The Treatment of Fractures of the Lower Jaw" (Part V. of *Contributions to Medicine and Surgery*), 13 pp. 5 plates.

1883. "Intestinal Disease and Obstruction" (Part I. of *Contributions to Medicine and Surgery*), 284 pp. + xvi, 5 plates, Appendix.

1883. "Nerve Inhibition and its Relation to the Practice of Medicine" (Part VIII. of *Contributions to Medicine and Surgery*), 47 pp.

1883. "Principles of the Treatment of Diseased Joints" (Part II. of *Contributions to Medicine and Surgery*), 151 pp.

1885. "The Collegian of 1666 and the Collegians of 1885; or, 'What is Recognised Treatment?'" (Part IV. of *Contributions to Medicine and Surgery*), 147 pp. First Edition, 1885; Second Edition, 1888.

1886. "The Principles of the Treatment of Fractures and Dislocations" (Part VI. of *Contributions to Medicine and Surgery*), 104 pp. 17 plates.

1887. "Fractures, Dislocations, and Diseases and Deformities of the Bones of the Trunk and Upper Extremities" (Part III. of *Contributions to Medicine and Surgery*), 127 pp. 16 plates.

1888. "A New Lithotomy Operation," *Provincial Med. Journ.*, April 1888.

1889. "An Argument with the Censor at St. Luke's Hospital, New York" (Part VIII. of *Contributions to Medicine and Surgery*), 45 pp.

1890. "Lithotomy" (Part VIII. of *Contributions to Medicine and Surgery*), 16 pp.

1890. "Fractures, Dislocations, Deformities, and Diseases of the Lower Extremities" (Part VII. of *Contributions to Medicine and Surgery*), 320 pp. 117 figs.

"Spinal Deformities." Projected in 1878 as Part IX. of *Contributions to Medicine and Surgery*. Never published.

Manual of Orthopædic Surgery. Announced in 1878 as shortly to be published. Never published.

WILLIAM JOHN LITTLE, 1887.
BORN 1810; DIED 1894.
A pioneer of Orthopædic Surgery.

CHAPTER IV

THE INTRODUCTION OF TENOTOMY

In this chapter a sketch is given of the work of three men—William John Little, George Frederick Louis Stromeyer, and William Adams. All of them had accomplished their best work before the subject of the previous chapter had put up his plate in Liverpool. All three men were prolific writers. For a full list of their publications the reader is referred to the Catalogue of the Library of the Surgeon-General of the U.S. Army.

IN three former chapters we have considered the principles and practice of three outstanding medical men—Hunter, Hilton, and Thomas—gathering from their life's work, as best we could, the facts which are useful for our present purpose—the repair of disabled limbs. We have tried to follow the manner in which they acquired their knowledge and the use they made of it in their modes of treatment. My object in this chapter is not to follow a man, but a movement—a movement which led surgeons to practise the section of tendons for the relief and rectification of deformities. We have seen that Hunter had carried out a series of experiments to study the effects which follow section of a tendon and the exact manner in which the breach so caused underwent repair. The events, however, which led to the introduction of tenotomy as a measure of surgical routine had nothing to do with Hunter's experiments, they occurred nearly thirty years after he was dead. The events we are to describe belong to the earlier half of the Victorian Era, for it was in 1837, the year in which Queen Victoria ascended the throne, that the first operation for the cure of club-foot by the section of tendons was carried out in England.

The man who can best serve as a guide, as we follow the march of events, is William John Little, the son of parents living in comfortable circumstances in the East End of London. Little was born in 1810, and was therefore a contemporary of Hilton. In his infancy he suffered from a fever which left him, when two years of age, with a palsied and deformed foot. As he grew up, so he informs us, his left foot became inverted, the heel became drawn up and a typical talipes equino-varus developed. At the age of 16 he was apprenticed to a neighbouring apothecary, but two years later, in 1828, his indenture being cancelled, he began the study of Medicine at the London Hospital, situated then almost in the open country but now buried in the great "East End." He entered his studies with the fixed intention of discovering what could be done for the relief of such a condition as he suffered from. He found that club-foot was regarded as lying outside the legitimate scope of surgery, and, in the opinion of his teachers, was properly confided, as his own case had been, to the care of bone-setters and sprain rubbers, who treated the condition with manipulations or instruments, often with a fair degree of success. A man, whose chief object in devoting himself to Medicine, had been the alleviation of a personal infirmity—one with which his senior contemporaries Lord Byron and Sir Walter Scott were afflicted—was not likely to remain content with the comfortless promise of his time. Hence we find him, all through his time of study at the London Hospital, seeking every opportunity of making himself acquainted with the actual condition of parts in deformed feet. Fortunately, for the aim he had in view, he possessed a facile command of the French tongue, and during his studenthood followed medical progress in the literature of Paris as well as in that of London. He became particularly interested in the work of the French surgeon Delpech of Montpellier—who was afterwards to fall a victim to a patient on whom he is said to have performed an operation for varicocele. Delpech had proposed and carried out in 1816, and again in 1823, section of the tendo Achillis for the cure of club-foot. He even approached Delpech with a view of having the operation carried out in his own case. The great French surgeon advised him to have nothing done—for several reasons, the chief being the risk of suppuration and sepsis. Thus as a student Little had his attention drawn to the possibilities of tenotomy. We see from his interest in the medical literature of foreign countries that Little was not a student of the ordinary kind, and we have further evidence of his eager search for knowledge when, in his twenty-second year, having passed through the curriculum of his own hospital, he went to study in the post-mortem room at Guy's Hospital, under Thomas Hodgkin—one of the most outstanding of medical men then in England—and to University College to study comparative anatomy under Dr. Robert Grant. In 1832, being then a young man of 22, and a member of the Royal College of Surgeons, he fixed his plate to a door in Billiter Street and set out to build up a practice in the City of London. The teaching of comparative anatomy was then finding favour in the medical schools attached to the hospitals of London, and Little became attached to his old hospital, the "London," as lecturer on this subject—a source of honour rather than of income.

When he put up his plate in Billiter Street, Little's ambition was to become a member of the surgical staff of his hospital and to devote himself to the practice of surgery—not of any particular branch, for specialism in surgery was then unknown. Two years of waiting saw his ambition wrecked. There was then a vacancy in the surgical staff, but his claims were passed over, and hence he had to change his plan of life. It is that change of plan which brings him into the history of orthopædic surgery. He resolved, as was not unusual with men so circumstanced, to become a licentiate of the College of Physicians of London and devote himself to pure Medicine. To become a licentiate it was necessary for him to spend two years at a University. He chose Berlin. He was drawn thither by the great and growing fame of Johannes Müller.

In 1834 we see this lame and somewhat sensitive Englishman set out for Berlin, armed with a letter of introduction to Müller from Grant of University College, and supported by the status accorded to him because of his office of lecturer in Comparative Anatomy at the London Hospital. When Little entered Müller's laboratory, he found there Schwann, Henle, Remak, and the other young men who, a few years later, were to reveal the cellular constitution of living matter. He had stepped into the centre of the most productive and progressive medical movement then in Europe. Little had every opportunity given him in Müller's laboratory of continuing his dissections of deformed feet. The condition revealed by his dissections supported the conclusions he had drawn from his investigations in England, namely, that surgeons were in error in believing club-foot was a result of an inherent defect in the growth of the bones of the foot. Little realised that the cause of the deformity lay in the soft parts—particularly in a disordered action of the muscles. With that conclusion Müller agreed. He also concurred with Little in regarding the condition as one which should be amenable to surgical treatment.

Before leaving England, Little had read of a young surgeon at Hanover, Stromeyer by name, who had modified Delpech's operation and was cutting the tendo Achillis for the rectification of club-foot. Müller agreed with Little, in face of an opposite opinion expressed by surgeons in England,

France, and Germany, that such an operation had a rational basis. Hence we find Little, in the summer of the second yeai of his study in Berlin, and the twenty-sixth year of his age, visiting Stromeyer in Hanover, to see with his own eyes how far such a condition as he himself suffered from could be relieved by the practice of cutting tendons.

GEORGE FREDERICK LOUIS STROMEYER,
BORN 1804; DIED 1876.

In 1831, at the age of 27, he introduced the method of subcutaneous operations to surgery and applied them in the treatment of deformities. By 1836 he had operated on 350 cases of club-foot, 200 cases of strabismus, and 120 cases of wry-neck. In 1838, while still at Hanover, he published his *Beiträge fur Operativen Orthopædik.* He was called to fill the chair of Surgery in several German universities before he finally settled as Professor of Surgery in the University of Keil. In 1854 he retired from that chair, and was succeeded by his son-in-law, Esmarch. He was a popular figure at medical meetings in England, and had a particularly warm reception during his final visit in 1872.

Louis Stromeyer was only six years Little's senior; he was born in Hanover in 1804, the son of a surgeon of that town who had strong leanings towards the art as practised in

England, and hence, when Louis finished his medical curriculum, his father recommended a sojourn in London to watch the practice of English surgeons. After spending almost a year in England, young Stromeyer commenced practice in his native town in 1828. We must look into his career during the eight years which elapsed between the time he began practice and the summer of 1836, when Little occupied a bed in his modest private hospital. It is worth our pains, for we shall see from how small a beginning a new movement will arise and spread to the utmost limits of the surgical world. From the outset of his practice Stromeyer applied himself to the treatment of physical disabilites and deformities. He fitted out a small private hospital, but found the establishment of the kind of practice he desired an uphill task. In 1831, when he was in his twenty-seventh year, the opportunity came, and he took it. The son of a local schoolmaster, a boy of 14, was the subject of club-foot—intractible, painful, the despair of his relatives and medical attendants. Stromeyer gave the lad a bed in his hospital, and settled down to devote eighteen months of unremitting attention and care to his case. As a last resort, he cut the tendo Achillis by a new—a subcutaneous—method; he found that the foot could then be flexed (dorsi-flexed), and that the cut ends of the tendon separated for three-quarters of an inch. The gap evidently frightened him somewhat; at least he again extended (plantar-flexed) the foot until the ends were in apposition, before he fixed it in a splint. At the end of six days he found on bending the foot that the cut ends of the tendon did not separate, but were so far united that they moved together. He gradually dorsi-flexed the foot, with the object of stretching the scar in the tendon. After practising these movements for a period of eight weeks, the heel was brought down to its proper level, and all trace of the equinus deformity removed. Stromeyer ascribed his success, not to the eighteen months of unremitting attention he had given to the case, but to the operation he had thus introduced into surgery. His was a mind which glowed with the enthusiasm of the missionary and philanthropist. In his eyes, the numerous cases of misshapen bodies and of deformed limbs which were to be seen on the streets of all great cities were a reproach to surgeons; they proclaimed the incompetency of their vaunted art. Tenotomy, he believed, would bring hope and healing to a neglected class of unfortunates and remove a slur from the practice of surgery. At the lowest estimate, tenotomy, he declared, reduced the time necessary for the cure of club-foot from months to weeks. One great truth he clearly recognised—one which has still to be insisted on—that the theory of congenitalism, which had been so consistently applied to explain the existence of a certain class of deformities, was a hindrance and curse to surgical progress; "congenital" was a shibboleth used to cover ignorance, and when applied to club-foot prevented surgeons from taking the first step essential for successful treatment—namely, an inquiry into the origin and nature of the condition. Nature, he said, could not cure club-foot; all its efforts could only make the condition worse. Heat and rubbing, he found, were useless in mending spastic or contracted muscles. He came to the conclusion that deformities were primarily due to disordered muscular action. To cut the tendon of a spastic or contracted muscle was therefore a rational means of treatment—the operation at once threw the disordered agent out of work.

Stromeyer was also impressed by the discoveries of Charles Bell. In 1836, the year in which Little came under his care, Stromeyer published a paper on lateral curvature of the spine, ascribing the deformity to a disordered action of the muscles of inspiration. Section of a tendon, he believed, not only relieved the tension of a muscle, but also, as Hunter had concluded, altered its functional behaviour. When a tendon of a spastically contracted muscle was cut, he observed that it passed into a condition of rest. Tenotomy he regarded as a means of giving rest to an inflamed joint. He saw how tenotomy could be profitably applied to defects and deformities of many regions of the body; he noted that squint was due to a disordered action of the ocular muscles, and introduced tenotomy as a cure for strabismus. Tenotomy was not only a cure for deformities;

it gave the surgeon an opportunity of directly affecting the disordered actions of muscles. Flat-foot, however, he regarded as primarily due to a weakness of the plantar and tarsal ligaments combined with a state of contraction in the muscles situated in the calf of the leg. He noted, in those who became the subject of this grave disability, that the feet and legs were cold, with a tendency to sweat and often marked by a deficient circulation.

Such was the condition of matters when Little entered the modest orthopædic hospital at Hanover in the July of 1836. There he had his club-foot rectified. In Stromeyer's notes we find him described as a heavily built man, walking on the outer side of his deformed foot. The results of the operation and of the after-treatment were highly satisfactory. When recovered, Little stayed on at Hanover, and was given numerous opportunities of perfecting himself in the technique of the new operation; he returned to Berlin cured and convinced that a new era had dawned for the deformed. He showed himself to Müller and to Müller's colleague, Dieffenbach, the surgeon; they were amazed at the success of the Stromeyerian methods. Dieffenbach put them in practice almost immediately; in the course of a little more than a year, he had operated on 140 cases of club-foot. Dieffenbach's method differed in one important respect from that used by Stromeyer. The latter, we have seen, first performed tenotomy, and then stage by stage and week by week rectified the deformity by manipulation. Dieffenbach, on the other hand, performed tenotomy and rectification at the same sitting. Dieffenbach was the advocate of force, Stromeyer of persuasion. In his practice, Little was a follower of Stromeyer.

Little, having read his thesis for the Doctorate of Berlin University, on "The Nature and Treatment of Club-foot," the first account published of the results obtained by the Stromeyerian method, returned to London early in 1837 and at once settled down again to practice in the City—his attention being directed in particular to the treatment of cases of club-foot. On 20th February of that year, he performed the first operation in England[1] for the relief of club-foot by section of tendons. Not only did he cut the tendo Achillis, as Stromeyer had done, but also the tendons of the tibialis posticus and flexor longus hallucis, because in his dissections of club-foot he had observed that much of the deformity was due to their state of over-contraction. His enthusiasm compelled the attention of his British colleagues; some were interested, others were sceptical, many were actively opposed to the new measure. Although of a retiring, modest nature, he proved to be the right man to lead a crusade. In 1838, with the help of relatives and friends, he succeeded in establishing the Orthopædic Institution—afterwards the Royal Orthopædic Hospital—the first of our British public charities for the relief of the maimed and deformed poor.[2] Into the labours of that institution he threw his full strength. In 1839 he published a treatise on *The Nature of Club-foot and Analogous Distortions*—dedicating the work to Sir Astley Cooper. He gave courses of lectures to students on orthopædic treatment of deformities; he published the course he gave in 1843–44 under the title of *The Nature and Treatment of the Deformities of the Human Frame.*

There can be no doubt that Little was the pioneer of orthopædic surgery in England. He regarded subcutaneous tenotomy as a great discovery—a surgical revolution. If in this he was too sanguine, he at least focused attention on the treatment of deformities, and particularly on muscles and tendons. We have evidence that his influence and labours made themselves felt both in London and Edinburgh. In 1839, James Paget, then a young man of 25, waiting at St. Bartholomew's Hospital for an appointment on the staff, was moved to investigate the repair and blood supply of tendons. He found that tendons were provided with a double supply

[1] He believed it was the first, until some time afterwards, when he discovered that Mr. Whipple, a surgeon at Plymouth, had employed tenotomy in the previous year (1836).

[2] It was at this hospital, in 1865, that his third son, Louis Stromeyer Little performed subcutaneous osteotomy for the correction of knock-knee —the first operation of its kind done in England. His son, Dr. E. Muirhead Little, is now senior surgeon to the Royal Orthopædic Hospital.

(1) from the arteries of the muscle; (2) from the arteries of their sheath. If a tendon were cut, he observed that both of these supplies took part in furnishing the "callus" of repair with nourishment. In Edinburgh, Syme also began to practise the operation of tenotomy. He cut tendons in deformed feet, forcibly rectified the deformity, and almost immediately turned patients out of hospital to let Nature complete the cure; we know now that in such cases she performs her part very ill. In Little, as in Stromeyer, and particularly so in the case of Thomas, we find a clear recognition that the essential part of the cure is not the cutting but the after-care of the case.

Little's desire to serve the cause of orthopædic surgery was restrained by the fact that his real aim in life was to become a physician. We have seen that he went to Berlin with a view of becoming a licentiate of the College of Physicians, and of devoting his life to Medicine. On his return to England he became a licentiate of the College, and in 1840 was appointed assistant physician to the London Hospital, and at a later date was appointed full physician and lecturer on Medicine in the College attached to the Hospital. We can well understand that the College of Physicians could not look upon his surgical endeavours with a favourable eye when considering the election of its Fellows, and hence it came about that Little saw young men pass him by while he had to wait on the door-step of his college; he was a man of 67 before he was made a Fellow. His interest in deformities, however, had its compensations; it drew him into the study of the disordered action of muscles, and led to his recognition and description of that curious form of infantile spastic paralysis known as Little's Disease. His belief in the efficacy of tenotomy ultimately waned; he came to think that a muscle which had been stretched without a previous tenotomy recovered better than one which had been stretched after its tendon had been cut. We find him in 1876, when he gave an address in Edinburgh, towards the end of his professional life, using expressions which show us that his early enthusiasm for tenotomy had become tempered by an experience of forty years. He then realised that tenotomy might be a curse as well as a benefit. Tenotomy applied as the sole measure for the treatment of deformities is certain to yield disastrous results. He came to the final conclusion, the same conclusion as we have seen H. O. Thomas came to, that it was the continuous care of the surgeon, the nursing and coaxing of the parts day by day, with an infinite expenditure of patience, which gave restoration of shape and function to deformed parts. In 1884, William John Little withdrew from active practice to live a quiet and retired life at West Malling, Kent, where he died in 1894. Although a physician, he was also the pioneer of orthopædic surgery in Britain. He did more than anyone of his time to rescue the maimed and deformed from the hands of a class of practitioners who, if some were skilled, most were ignorant and unscrupulous. Stromeyer's career had come to an end long before; he died in 1878, at the age of 74.

To ascertain the principles which guided the practice of the men who followed in the orthopædic movement started in England by Little, we shall follow the career of William Adams. When Little returned from Berlin in 1837, Adams, a lad of 17, had just been apprenticed to his father, a surgeon in Finsbury Square, in the City of London. He joined St. Thomas's Hospital as a student, working under Hodgkin and Green, and after a course of four years became a member of the Royal College of Surgeons, England, in 1842. By that time the orthopædic movement started by Little was well afoot. At Hodgkin's suggestion Adams became curator of the Museum at St. Thomas's Hospital, and worked there at pathology, waiting for an appointment on the staff, and trying, while he waited, to build up a practice in the City. In 1851, being then 31 years of age, and seeing no hope of an appointment to his own hospital, he sought and obtained a place on the staff of the orthopædic hospital founded by Little. We can see by the way he then set to work that he knew how scientific surgery should be built up. There were still uncertainties as to the exact manner in which repair was effected after a tendon is cut. Adams carried out a series of experiments on sixteen rabbits, cutting the tendo Achillis, and studying closely each stage in the process of repair. His description of the "nucleated blastema"—a phrase which shows us that the period of the microscope had dawned—need not detain us, nor need we linger over his insistence on the part taken by the sheath or peritendineum in supplying the material needed to make good the gap in a cut tendon. If time permitted, it would repay us to review the results of his dissections of deformed feet in which the operation of tenotomy had been carried out at varying periods before death had occurred from some accidental cause. His descriptions of the condition of the parts to be seen in the dissections of deformed feet are still the best to be found in our libraries. In one case of tenotomy of the tendo Achillis, he demonstrated that $2\frac{1}{4}$ inches of new tendinous material had become inserted in the course of repair; in another foot, where the tibialis posticus had been cut within its synovial sheath behind the malleolus, he observed that repair had failed and at the point of section—behind the internal malleolus—the tendon had become adherent to the bone. He noted that in every case of tenotomy the tendon became adherent to its sheath at the point of section, the bond being widest and firmest on the deep aspect of the tendon. Adams availed himself of every opportunity to look beneath the surface and obtain an accurate knowledge of the condition of the deep parts—bones, joints, ligaments, and muscles, which had to be dealt with in the rectification of deformities. He investigated particularly the condition revealed by the dissection of club-foot, and in 1864 was awarded the Jacksonian Prize of the Royal College of Surgeons for his dissections and investigations.

In the Museum of the College we may still study the club-feet which he made the subject of his well-known treatise. When, however, we seek for a deeper knowledge of the cause and prevention of deformities, and of the behaviour and action of muscles and of nerves, we shall search Adams's publications in vain. He failed to realize, as was the case with so many of his contemporaries, that physiology has a direct bearing on surgery. With Stromeyer and Little, he believed that deformities should be slowly reduced after the necessary tenotomies had been performed, and that rectification had to be effected by the application of rigid machines. But he differed from Little and agreed with Dieffenbach in believing that, in congenital cases of club-foot, treatment could not be commenced too early. Little preferred to wait until the child was old enough to learn to walk.

In 1871 he introduced a subcutaneous operation for the relief of anchylosis of the hip-joint. We see in that operation the application of the subcutaneous method to osteotomy. He there applied to the hip-joint a method which had sprung out of the practice introduced by Stromeyer for section of tendons. Subcutaneous osteotomy, however, had been practised in Germany long before the date at which Adams applied this method to the hip-joint. Adams was a kind and pleasant man and, like his contemporary Little, lived to an old age, dying in 1900 at the age of 80. Both had made a reputation as orthopædic surgeons before Thomas's sun had risen. Indeed, in the year 1857, when Thomas was spending his final year of study at University College, London, Adams published his work on the *Principles and Practice of Subcutaneous Surgery.*

The Law of Ligament.—When we examine the principles and practice of Stromeyer, Little, and Adams, all of them pioneers in orthopædic surgery, we are struck by the importance they attach to ligaments in the production and treatment of deformity; ligaments seemed to them almost as important as muscles. Even at the present time the essential function of ligaments is often misunderstood, and so long as this is the case we cannot hope to effect an object which is quite as important as the rectification of deformities—namely, their prevention. Hunter's teaching as regards the respective functions of muscles and ligaments in the mechanism of the human body is very definite. Muscle is the only form of living tissue which can be applied for the continued support of parts without undergoing a passive stretching or elongation. A ligament cannot perform such a function because it is composed of living cells possessing no power of contraction, no power to shorten as their load increases, such as muscles are endowed with. When the living cells and fibres of a ligament are subjected to a *continued* burden they invariably elongate. Hence Nature never uses ligaments either for the purpose of passive support or of active maintenance of parts in position; she uses them only as safeguards—as the means of limiting the movements of joints. They come into play chiefly when the muscles which guard and surround a joint are forced beyond the compass of their normal stretch. Ligaments come into action only when the normal muscular support and defence of a joint has broken down.

This law can be best illustrated at the shoulder-joint. When the shoulder muscles are paralysed, or when they are thrown out of action by deep anæsthesia, the head of the humerus drops away from the glenoid cavity under the weight of the arm; the shoulder-joint can then be moved far beyond its normal radius of action. The ligaments thus become the sole agents by which movements are limited and checked. If extreme movements are continued the ligaments will become stretched. Or if in the dissecting room we strip the muscles from the shoulder and leave the humerus attached merely by its ligaments, we can see then that in all normal movements they never become taut until the usual limits are exceeded. The real ligaments of the shoulder-joint, as of every other joint in the body, are the active defensive contractile muscles.

Now man's upright position has made him more dependent on the ligamentous function of muscle than any other animal. His shoulders, when he stands or sits, have to be steadily supported by muscles; in the upright position every one of the twenty-four vertebræ of his back-bone has to be kept continuously balanced one upon the other; the contents of his abdomen have to be constantly supported and braced by the contraction of the muscles of the abdominal wall and thus prevented from prolapse towards the pelvis. Ligaments are useless for such purposes; Nature never employs them for such ends, because under conditions of this kind they must stretch. We see the same principle applied in the maintenance of the joints of the lower extremities. We cannot stand without the muscular braces of our hip, knee, and ankle joints coming into continuous action. At the hip-joint it is true that, when the joint is fully extended, the action of the flexor muscles is braced by the strength of the great capsular ligament, but, however taut that ligament may be, the ilio-psoas muscle can also be felt in action. At the knee, ligaments limit and control the articular surfaces at all stages of movement, but at no stage are the ligaments left without the active support of muscular action. It is easy to demonstrate that the maintenance of the plantar arch owes nothing to ligaments; that can be demonstrated in the living foot and leg, and also in the dissected parts. It is quite clear that ligaments are passive parts; their elongation is not a cause, but a consequence of the deformity. In short, in all static deformities of the human body the cause has to be sought for, not in ligamentous changes, but in the disordered action of the muscles, and we shall never succeed in preventing or mending static deformities until the truth of this law of the function of ligaments is clearly realised. That will become very apparent when we come to examine the conditions under which static deformities of the body arise.

MARSHALL HALL,

BORN 1790; DIED 1857, aged 67.

From a photograph taken about 1850, when he was 60 years of age.

CHAPTER V

MARSHALL HALL AND THE BEARING OF HIS DISCOVERIES ON ORTHOPÆDIC PRACTICE

Marshall Hall was born at Basford, near Nottingham, 18th February 1790, and died, when he had retired from London to Brighton, on 11th August 1857. He was thus senior to W. J. Little by twenty years. His professional life may be divided into the following four stages:—

(1) From 1805, when, at the age of 15, he was apprenticed to a chemist in Newark, until 1817, when, at the age of 27, he finally settled to medical practice in Nottingham. Of the twelve years included in this period, four were given to his apprenticeship, five (1809–1814) to medical study and clinical work in the University of Edinburgh, where he had a brilliant career, and three years to settling down to practice; his first choice—Bridgwater—failing to satisfy his ambition.

(2) A Nottingham stage—from 1817 to 1826. In the latter year he

suddenly set out for London, aged 36. This stage is marked by activity in publication, in experiment, and in practice. His clinical observations were chiefly directed towards the elucidation of problems relating to certain diseases peculiar to women.

(3) An earlier London stage extending from 1826, when he settled to practice and experiment at 17 Keppel Street, Bloomsbury, until the publication of his discovery of the "Reflex Function of the Medulla Oblongata and Medulla Spinalis" in 1833.

(4) A later London stage—from 1833 to 1857, a period of twenty-four years—chiefly occupied with the application of his scientific discoveries to clinical conditions.[1]

THE men whose orthopædic practice we have followed hitherto (Hunter, Hilton, Thomas, Stromeyer, Little, and Adams) fixed their attention on the external parts—on the muscles, bones, and joints. They were well aware that deformities were produced by a disordered action of the muscles, but of the nature of that disorder they had no clear conception, because they had a very imperfect picture of the elaborate machinery of nerve cells and nerve fibres which controls and co-ordinates the workings of the muscular system. In the minds of all of them—even of the later members of the group—the spinal cord was little more than a mere cable of nerve strands linking the brain to the body. From 1811, when Hilton was still a boy at school in Castle Hedingham, until the autumn of 1832, when Marshall Hall thrust his theory of reflex action on the jealous and sceptical medical public of London, Charles Bell had been teaching in London his "New Idea of the Anatomy of the Brain." Bell had shown that every muscle was supplied with two sets of nerve fibres—the one to link the brain to the muscles, and the other the muscles to the brain. He was the first to explain the form and arrangements of the parts of the central nervous system on a physiological basis; the anterior parts of the spinal cord he regarded as devoted to the transmission of messages sent out by the brain to the body, while its posterior parts were concerned with messages passing towards the brain from the

[1] For a fuller account of Marshall Hall's life the reader is referred to *Memoirs of Marshall Hall*, by his widow (London, 1861); and for the best account of his researches on the nervous system to the *New Memoir* (1843).

body. In short, Charles Bell conceived that all the active machinery of the central nervous system was situated in the brain itself and that the spinal cord was a mere annexe used for purposes of transit. It was Marshall Hall's outstanding service to demonstrate conclusively that what was supposed to be a mere corridor for brain messages was in reality one of the chief workshops of the human body—a drawn-out laboratory which was staffed and stocked with wonderful mechanisms for regulating the behaviour and action of muscles.

Marshall Hall was a physician; but the effect of his discovery extended far beyond the realm of pure Medicine; it percolated into every realm of biological thought; it gave a rational interpretation to myriads of clinical and functional manifestations which medical men had either puzzled over in vain, or passed by; it gave a clue to correct diagnosis and a basis for rational treatment. Above all, as we shall proceed to show, Marshall Hall's discovery is one of the first magnitude for the orthopædic surgeon.

When Marshall Hall first announced his discovery at a November meeting of the Zoological Society of London in 1832, he was a man of 42, with a house in Manchester Square, a carriage and pair and a consulting practice which brought him over £2000 a year. He had not been educated in London; he had no appointment on a hospital staff; he had no social influence. We want to see how this short, stoutish, plain man, with a homely, thoughtful countenance, and an alert, active, decided step and manner, self-contained and somewhat self-centred, had elbowed his way into the medical and scientific centres of London; in particular, we want to see how he came to make discoveries which had a revolutionary effect on medical practice. We shall find that his early career is one of the most chequered and instructive to be found in medical biography. The path he followed differed from that which guided Hunter, Hilton, and Thomas to their life's goal. From the day he commenced the study of Medicine in the University of Edinburgh, he became the typical, hard-reading, successful student, verifying and amplifying his reading by close and exact observation by the bedside. He remained a hard reader and active writer until just prior to the point at which we have taken up the thread of his life, when, for a definite reason, he purposely adopted the experimental method of gathering knowledge, and at one bound secured for himself a permanent place in the history of Medicine.

The date of his birth takes us to 1790—three years before Hunter's death. He was born of a Saxon Midland stock—with an infusion of that hybrid strain which reached England through Normandy. His father was a cotton spinner and bleacher near Nottingham, the first in England to use chlorine for bleaching cotton. His family were seriously and conscientiously minded, religious, interested in chemical problems and inventions. His brother Samuel was a prolific and noted inventor. Marshall, at the age of 15, was apprenticed to a chemist or druggist in Newark, but in 1809, four years later, we find him in the dissecting room of Edinburgh, the first to arrive in the morning and the last to leave at night. He opened his student's career in Edinburgh by contributing to *Nicholson's Journal* a research in chemistry; in his third year his fellow-students elected him President of the Royal Medical Society. We find him using spare intervals to master the Latin and French tongues. Altogether he spent five years in Edinburgh, three as a student living on £55 a year, allowed to him by his father, two as a resident in the Infirmary, where he was furnished with board and £20 a year. Leaving Edinburgh in 1814, in his twenty-fourth year, with a great reputation and the manuscript of a work on the "Diagnosis of Diseases" in his valise, he reached London, where he called on Dr. Matthew Baillie, John Hunter's nephew, and then set out for Paris in company with an invalid gentleman. On his return, seeing no prospect of an opening in London,—a youthful ambition,—he settled in Bridgwater in the year of Waterloo. After waiting there for two years for a kind of practice which never came, he borrowed £100 and settled in Nottingham, marking his arrival there in 1817 by publishing his book on *Diagnosis of Diseases*. A copy of this book, bestowed on Dr. Matthew Baillie, bore fruit; the London consultant recommended the young practitioner to the Manners

6

family, which opened for him a lucrative practice among the county people of the Midlands. He settled himself exclusively to his studies, books, and practice; he began to write and publish on the most diverse subjects—scientific, as well as medical. Some of his papers were published in the *Journal* of the Royal Institution, others in the Medical Press, and some in book form. We note particularly that he devoted much attention to puerperal conditions—as if it were his intention, should an opportunity present itself, to specialise in the diseases peculiar to women. The student-practitioner prospered exceedingly; there was a quality in him—perhaps it was an interest centred in his own particular affairs—which made him rather an Ishmaelite among his medical colleagues of the county town. Nevertheless in 1825, when he had been in Nottingham for eight years, he was appointed physician to the hospital there.

On an August morning of the following year, this successful bachelor of 36 set out for what was to all outward appearance merely a temporary visit to London; but, arriving there, he established himself at No. 17 Keppel Street, Bloomsbury, and gave orders to have his home in Nottingham dismantled. He was a man with business-like instincts and understood public susceptibilities. He made an appearance abroad in his carriage and pair; he counted on retaining the medical patronage of Nottingham families, and in that he was not disappointed. In his first year in London he earned £800, which gradually rose until it exceeded £4000 per annum. In establishing himself as a consultant in London he was fulfilling one ambition, but there was another very near his heart; he wished to become a Fellow of the Royal Society. We have seen that during his life at Nottingham he had devoted himself not only to many medical subjects, but also to problems of a physical nature. When he came to London he was particularly interested in the effects which followed blood-letting, and saw there were many points connected with the circulation of the blood which required further investigation. It also became clear to him that if he were to further his candidature for the Royal Society he must not depend on the experiments produced in the bodies of his patients by disease, but must himself elicit such information as he required by direct experimentation on animals. Hence we find him setting aside a room in his Keppel Street home as a laboratory, installing the most improved microscopes and an ample supply of frogs, newts, salamanders, and turtles to supply the living lungs, mesenteries, skin, and webs which were to give him an opportunity of studying directly the manner in which living tissues were furnished with blood.

His researches had reached a point in 1831, by which time he had married and taken up a new abode and a new laboratory at No. 14 Manchester Square, when he thought them ripe for publication. His paper was read at the Royal Society, but was refused a place in its *Transactions* ; yet in this paper there was a big and important truth, namely, that the capillaries—"methemata" he called them—were the essential element of the circulatory apparatus; they were the tissue feeders; arteries and veins were merely necessary adjuncts.

It was in the course of these experiments on the circulation in capillaries that he made a chance observation which directed his attention to the nature of the functions carried on in the spinal cord. He had cut off the tail of a salamander, and happening to touch the skin, observed that there was a quick movement in the amputated tail. That response would probably not have excited the curiosity of a hundred ordinary observers; it was just what they would have expected to happen, and what Marshall Hall's predecessors had often seen happen. It immediately excited his attention because it was taught in his time that a muscle could be stimulated to action in only one of three ways: either by touching or stimulating it directly, or by stimulating its nerve, or by a volitional stimulus from the brain. Now he knew that the movement he had evoked in the salamander's tail was due to none of these three kinds of stimulation; there must be a fourth way of calling muscles into action, and he, as was his wont when he encountered exceptional manifestations, followed up the clue. He knew, as only born investigators know, that the unexplained, ordinary occurrence and the exceptional case are the sure clues to discovery. The eye that interprets the falling apple always works from a brain that is fit to grasp the law of gravitation. It was so with Marshall Hall; he immediately set out to discover the significance of the response he had elicited. He destroyed the part of the spinal cord which lay in the detached tail, and found, on touching the skin, no movement could then be elicited by that method of stimulation. He thus was led to suspect that in the spinal cord there resides a power, centre, or mechanism, for the production of such movements as he had witnessed in the salamander's tail. He immediately set to work to see whether his guess was right. He took a snake and cut its spinal cord across, just behind the head—he had thus separated its brain, the recognised centre of muscular machinery, from the body. The snake remained perfectly still until he touched it, or a puff of wind swept over it, or until the table was struck on which it lay, and then the snake moved and kept moving until he wrapped it in cotton wool and thus brought it to a standstill by protecting it from all external stimuli; so soon as it was brought to a standstill and its skin again protected, all movements ceased. The brain was clearly the source of voluntary movements, and the spinal cord the seat of the no less wonderful manifestation, purposeful and yet involuntary movements. He had accomplished in the decerebrate snake that which Hilton wished afterwards to accomplish in a diseased joint. Marshall Hall had discovered a physiological method of giving perfect rest by the withdrawal of all external stimuli. With the destruction of the spinal cord all power of movement was destroyed. He then took a turtle and cut off its head. On touching the nose, which he knew from Bell's teaching to be a respiratory area, the floor of the mouth executed the respiratory movements which occur in one phase of breathing in turtles, but when he destroyed the medulla oblongata, he could no longer elicit the respiratory act. He concluded that when the nose was stimulated by touch, a message was carried to the medulla by the fifth nerve, and in a centre there excited another message which was *reflected* along the motor nerves controlling the muscles of the mouth and throat. The arc of nerves, with which we are now all familiar, found its first concrete realisation in the brain of Marshall Hall; he recognised that a stimulus was necessary to produce the reflected or reflex muscular act. In his pictured arc there were no nerve-cells; a score of years were to pass before the microscope revealed their existence in the spinal cord and other nerve centres. Every part of the body which has to be treated by the orthopædic surgeon is furnished with its reflex arcs.

The manner in which Marshall Hall pursued his clue marks him out as a born detective in the realms of biology. To test whether or not his system of arcs really existed, he proceeded to expose the spinal cord of the headless turtle; he touched the spinal cord with a needle; the limbs were thrown into movements; messages, he concluded, must have been excited at the point touched, and from there they had passed up and down the cord and been reflected to the muscles of the limbs. He touched the posterior nerve-roots to a fore and then to a hind limb; the limbs were thrown into movements; he exposed and touched nerves of the trunk; movements of the limbs and body followed. He touched the skin of the trunk and of the limbs, and with each prick found that the movements evoked by stimulation of the skin were more vigorous than those which followed stimulation of nerves, the roots, or even the cord itself; evidently the skin was the chief end-station for evoking such movements. All of these manifestations could be explained if it was assumed that there were centres in the spinal cord, and that such centres were linked to the body by reflex nerve arcs. He noted, in the headless turtle, that the sphincter of the anus retained its normal tonic power and function, but when he destroyed the hinder part of the spinal cord, all tone and contraction instantly disappeared from the sphincter. *The power or mechanism which regulated the tonus or function of a sphincter was clearly resident in the spinal cord.* On pricking the headless turtle he could evoke movements in the limbs; it was also clear that in the spinal cord there was resident a mechanism which could time and regulate the contraction of muscles. And lastly, and for our present purpose, he made a most important observation—when he removed the spinal cord from the headless turtle the limbs became limp, all tone and degree of contraction disappeared from the muscles of the limbs. The spinal cord, he concluded, regulated the balance which exists between the various groups of muscles in a limb. It gave each muscle its appropriate degree of tone. That passive contracture of muscles which can work such dire deformities in diseased joints was, on Marshall Hall's showing, to be traced to a disordered function of the spinal cord. The spinal cord regulated the balance of the limb muscles and determined the posture of joints.

Marshall Hall was duly elected a Fellow of the Royal Society in 1832, and on 20th June 1833 he read his paper to the Society, "On the Reflex Function of the Medulla Oblongata and Medulla Spinalis," in which he demonstrated the existence in these parts of "a principle of action not hitherto distinguished with sufficient precision." I do not think that at any time, either before or since, a discovery of such wide-reaching significance has ever been announced with such brevity, clearness, and completeness. We have to remember that in 1833 our knowledge of the finer structure of the central nervous system was crude and imperfect, yet he pictured an arc system and grasped the full bearing of his discovery on the realm of Medicine and Biology. He had an instinct for sign-post facts and critical experiments. A multitude of functional and clinical conditions, which his predecessors had noted and partly explained, such as coughing, sneezing, swallowing, defæcation, micturition, involuntary movements of limbs, etc., fell into their appropriate place in his system of "reflex arcs." He saw that the reflex function of the cord and medulla could be exalted—by opium in the frog and strychnine in the mammal; that the manifestations of tetanus and hydrophobia received a rational explanation for the first time. Asthma and epilepsy he also wished to bring within the scope of his reflex law. The excito-motor or reflex function of the spinal cord he found to be more easily excited in the young than in the old; if a teat touched the lips, or a finger pressed the palm, of a newly born child, it was grasped by lips or fingers. The anencephalic human fœtus—with its open, frog-like skull—he regarded as a purely reflex mechanism; it reacted only when stimuli were applied. He observed how all the entrances and passages at the front end of the body, all the muscles and operations connected with the entrance of food and breath, were presided over and manipulated by his systems of reflex arcs. The action of the canals and muscles at the hinder end of the body were regulated by a similar system; defæcation, micturition, parturition were reflexly controlled acts. He found he could elicit such acts by stimulating the mucous linings of these canals; tenesmus, strangury, renal and biliary colic, the localised contractions of the belly wall, and the manifestation of sympathetic pains which accompanied diseases or disorders of the intestinal tract, were all brought by him within the scope of his new law. He found that reflexes could be set up by stimulation of the dura mater, pleura, and peritoneum; the muscular disturbances which may attend the eruption of teeth or the presence of worms in the rectum could be explained on his newly discovered system of reflex arcs. His conclusions which most concern us now relate to the muscles of the limbs; they, too, were controlled by the action of stimuli speeding along his postulated system of reflex arcs. Stimuli which arose in the moving limb were carried to the spinal cord by "incident" or excitor nerves, and were there "reflected" along the out-going or motor nerves to the muscles. Each contact of the foot with the ground he supposed to call forth a new series of movements. At first he believed that the "tone" of muscles was also a true reflex manifestation, but later he regarded the tone of muscles as a function of the cord, manifested independently of the arrival of peripheral stimuli.

Marshall Hall leaves us in no doubt as to the basal principle which regulated his treatment of disease; he followed the same principles as had guided Hunter. As a physician he desired to be Nature's servant—a sincere and outspoken servant, if not a very humble one. "I am persuaded," he said, "that a knowledge of the healthy body action is the only foundation for practical medicine and the only remedy for quackery." "A mere practical man is a quack" was one of his aphorisms. He would have nothing to do with a policy of treatment that was destitute of a rational basis—however popular that policy might be. "Hydropathy" he rejected "because it was a system of gambling." Yet there never was a man so misconstrued as Marshall Hall by his contemporaries; the blame is not to be saddled altogether on his contemporaries; he had an unconscious art of rousing antagonism, jealousy, and mistrust. After he had made his justly celebrated communication to the Royal Society in 1833, that body turned its back on him, and in 1837 refused to publish his further inquiries into the nature and manifestation of reflex action. It also, most unjustly, refused his early paper on the nature of capillaries and of their circulation. He had to seek a means of publication abroad, first in Berlin, where Johannes Müller, who had helped W. J. Little, gave Marshall Hall's work a warm welcome, and afterwards in Paris, until 1850. At the latter date, when he was 60 years of age, a reconciliation was effected and he became a member of the Royal Society. He "was at a loss to understand the action of his critics," who made "numerous ill-natured and disreputable innuendoes," and "seek to deprive me, if possible, of the just rewards of my labours." "He was weary," he said, "of wordy discussion," and "appealed from the first half of the nineteenth century to the second." The second half returned an emphatic verdict in his favour. He earnestly desired, in his earlier time in London, to obtain a post at a teaching hospital, but not one of the recognised schools would open a door to him. His work had no better reception from the younger generation of academical scientists than from his consultant colleagues. An Edinburgh student—afterwards well known—of the name of E. B. Carpenter was given ample space in the pages of the *British and Foreign Medical Review* to adopt the arrogant and brutal methods then practised by the "Edinburgh Reviewers" to bludgeon the ablest man in British Medicine.

I cannot resist throwing a side-light on the attitude adopted by his London contemporaries at the time his reputation was being so much canvassed. He was particularly anxious to know what Johannes Müller had written about his discoveries, but could not read the German tongue. Dr. Budd told him of a young student of the name of James Paget, then in his second year of study at St. Bartholomew's Hospital, who could be of help to him. Marshall Hall availed himself of this student's assistance, for, on 29th November 1836, James Paget, being then 18 years of age and Marshall Hall 46, wrote in a letter to his father, who lived in Yarmouth: "I have made the acquaintance of Dr. Marshall Hall lately; Budd introduced me to him, as able to tell him what Müller said of his reflex functions. This I have done and he has sent me copies of his original paper in the *Phil. Trans.* and his lectures on the Nervous System. He is certainly a sharp fellow, but I should think rather a monomaniac on the *reflections*." [1]

When we see how one of the best of the young intellects in England dismisses rather airily the labours and discoveries of Marshall Hall, we can hardly wonder that John Hilton preferred to think of nerve trunks than of spinal centres as the functional elements which presided over the muscular system of the body. Neither he nor the line of surgeons which followed him perceived that Marshall Hall had given a new basis for the diagnosis of all disordered actions of muscles and a sure principle on which a rational means of treatment could be based for the recovery of deformed and disabled limbs. Nay, I would go further; the full bearings of Marshall Hall's discoveries, and particularly the significance of the further investigations made by those who have so brilliantly explored the field which he had opened, have not yet been fully utilised by modern orthopædic surgeons. But of that I shall write more fully in a subsequent chapter: Marshall Hall made it impossible for us to think any longer in the terms of single muscles, bones, and joints; all of them, in the living state, are combined to form a functional whole by the spinal cord and its system of reflex arcs.

[1] *Memoirs and Letters of Sir James Paget.* Edited by Stephen Paget, 1901.

We cannot say good-bye to this "simple-speaking, plain-living" British physician without adding a postscript relating to the end of his career. In 1852, when he was in his sixty-second year, he became afflicted with a complaint, at first obscure, but afterwards proving to be a malignant growth in the commencement of the œsophagus, and withdrew to Brighton. His mind was as active as ever; it was in those last years that he worked out his postural method of artificial respiration—a method which is based on the elasticity of the thorax. The thorax is compressed, and air expelled from the lungs when the patient is turned face downwards; it expands, and air is drawn in when the body is turned on its side. Like his discovery of "reflex arcs," the elaboration of this method of artificial respiration was based on the experimental observations he made on the capillary system of newts, experiments undertaken to win an election to the Royal Society. His investigations led him on to study the circulation in the newt's lung; in that research questions relating to respiration arose, and the relationship of respiration to temperature and to the excitability of muscles. He had to appeal to comparative anatomy to solve his problems, and it was while in this pursuit that he made the discoveries which gave us a new light on the nervous system and a new method of recovering the apparently drowned. Up to his thirty-sixth year he used, with extraordinary industry, the accepted clinical methods of his time; he broke no new ground. But the moment he adopted the experimental method and applied it to lowly forms of life, where physiological problems are manifested in simpler terms than in man, he at one bound took a place in the front rank of makers of medicine. It was Marshall Hall who showed us that we may learn rational methods for the treatment of human limbs even from a newt's tail.

GUILLAUME BENJAMIN AMAND DUCHENNE,
BORN at Boulogne, 1806; DIED in Paris, 1875, aged 69.

CHAPTER VI

DUCHENNE OF BOULOGNE AS ORTHOPÆDIST

Summary of his Career.—Marshall Hall was already apprenticed as a chemist in Newark when Duchenne was born in Boulogne in 1806, also the year of Malgaigne's birth. Stromeyer was two years and Amédee Bonnet of Lyons four years his senior. Duchenne's career thus belongs to that period of the nineteenth century during which orthopædic surgery underwent its first great development. He was the son of a sea-captain, and received his preliminary education at Douai. In 1825, at the age of 19, he commenced his medical studies in Paris, reading his thesis and receiving his doctorate in 1831 at the age of 25. The teachers who most impressed him as a student were the physicians Louis and Chomel, and Magendie the physiologist. There was then a young but strong orthopædic school in Paris, led by Bouvier, but at this time neither the action of muscles nor the treatment of deformities had any attraction for Duchenne. That phase in his life began soon after his return to Boulogne, where he commenced practice in 1831. It was then he conceived the idea of using the Faradic current, at first for the treatment of chronic joint and muscular conditions, but afterwards for determining the use of muscles. In 1842, at the age of 36, after spending eleven years as a medical practitioner in his native town, he set out for Paris with a view of increasing his opportunities of acquiring knowledge rather than of adding to his income. At first his attention was directed to lesions of nerves and muscles; his study of orthopædic cases came later. In 1867, when he was 61 years of age, he published the work in which was summed up a lifetime of observation—*Physiologie des Mouvements*. In 1871, at the close of the Franco-Prussian War, he suffered an apoplectic seizure. His death in 1875 was scarcely noted in the Medical Press of the period.[1]

BEFORE setting out to survey the observations and discoveries of Duchenne of Boulogne, so far as they relate to the treatment of disabled muscles and joints, it is necessary for us to glance at what was happening in the Royal Institution, Albemarle Street, London, in 1824. Sir Humphry Davy presided over its destinies at that time; Michaël Faraday, his assistant, was making the discoveries which were to give the world its dynamo, and, as a minor issue, was to provide our friend Duchenne with the instrument by which he could dissect the living human body in a new way and earn for himself a permanent place among the benefactors of Medicine. It may serve to connect Duchenne with the men whose career we have just surveyed if I mention that, in the year 1824, Marshall Hall, who was almost of an age with Michael Faraday, and still in Nottingham, was contributing scientific papers to the *Journal* of the Royal Institution. Charles Bell was its frequent visitor; indeed it was a slighting remark of Sir Humphry Davy's, to the effect that the human body, so far as the anatomist was concerned, was an exhausted mine, which stimulated Bell to his discoveries. In the same year John Hilton joined the twin hospitals in the Borough as a medical student.

We propose to make Duchenne's acquaintance in the year 1835 as a medical practitioner in the seaport of Boulogne, when he was approaching his thirtieth year, for 1806 was the

[1] For an account of his life and works, see *Selections from the Clinical Works of Duchenne of Boulogne*, by G. V. Poore, New Sydenham Society Publications, 1883. See also a sketch of his life by Dr. A. Motet, *Annales Medico-Psychologiques*, 1896, 8th ser., vol. iii. p. 426.

year of his birth. He was therefore a contemporary of our Hilton, and not so unlike him in personal appearance; both were rather short, stoutish men with plain, intelligent faces. Little more than the Straits of Dover lie between their birth-places, for Duchenne was born in Boulogne, and retained all through life the provincial accent and deliberate speech of his native town. His father was a sea-captain, and his contemporaries have left it on record that, so far as personal appearance was concerned, he bred true to his stock. He had much of the Breton in him; that one can see in his widely set full jaws, his rounded head, his dramatic manner and excited speech. We shall find that this patient and retiring practitioner of Boulogne, if he had no great love for books, did possess an absolutely insatiable appetite for facts gathered at first hand. He had the greed of an ambitious capitalist; the more his account in the bank of knowledge grew, the more he longed to add to it; and yet it was a fortune which was always placed freely at the disposal of even the meanest medical beggar. He had a polemical side; like Hunter he was jealous of the title-deeds of his scientific property. He had his disputes *vive et passionée* with his contemporary Landry. "Landry," he said, "seems to think his mission in life is to contradict me." As for Remak of Berlin, it was beyond the power of man to get understanding knocked into him.

At the point we have taken up in Duchenne's career, he was in the fourth year of his practice, having returned to his native town in 1831 after spending five years as a medical student in Paris. As a young practitioner he had come heir to those chronic cases of rheumatic and indefinite pains which had been the despair and plague of his senior and successful colleagues. For such cases he had resorted to a new means of treatment—the application of the Faradic current by a method which was being used by Magendie in Paris in physiological experiments. The skin had to be punctured so that the induced or Faradic current might be applied directly to muscles and nerves, a rather painful method which naturally did not appeal to patients. Duchenne, therefore, sought for a more gentle method of applying the induced current for the treatment of nervous pains and muscular disorders, which he succeeded in doing by simply placing a wet sponge on the terminals of his battery. "I deem myself," he said, "the unquestionable inventor of this method of applying electricity." He also coined an apt term, viz. Faradisation, to designate the application of such induced currents to nerves and muscles. It was characteristic of the man that the curative effects of Faradisation were allowed to retire into the background; his interest became centred on the fact that he had elaborated a new method for the accurate study of the living muscles of the human body. He immediately set himself to stimulate each muscle and to observe its action, and to his surprise discovered that his ideal anatomist Winslow, from whom Hunter had drawn his original store of muscular knowledge, had left a field from which he could still glean a rich harvest of new and living facts.

Of the twelve years spent in Boulogne we know nothing; he made no announcement of any discovery during that time. All we can learn is, that in 1842, when almost 36 years of age, he set out for Paris carrying with him his beloved battery—the key which was to unlock for him the door to fame. Marshall Hall was exactly the same age (36) when he made his bolt to London, and both were faced with the same problem—how to establish themselves in the medical life of their respective capital cities. We have seen how Marshall Hall solved the problem, and we are now to see how Duchenne does the same. His ambitions lay in a different direction: he desired no fashionable home, only simple apartments; he preferred his jerky, alert walk to a carriage and pair; he did not wish to be trammelled by the responsibilities which encompass a staff appointment to a great hospital; all he desired was access to the clinical fields of Paris, and permission to glean, as best he could, behind the band of appointed harvesters.

You will see that we have in Duchenne one of the most remarkable characters in medical biography. He was Don

Quixote and Sancho Panza rolled into one. It must be apparent to every one versed in the privileges and prejudices of medical life that he had to face difficulties of various kinds. There was, in the first place, the bread-and-butter problem. As for that he had a modest competency, but he had also a practice, for in his published writings we find references to cases and fees which were placed in his way by Bouvier, the leading Parisian orthopædic surgeon of that day, by Nélaton, and by his old teachers, Louis and Chomel. His chief difficulties were of quite another kind. How was he to obtain the consent of hospital patients over whom he had no control, to apply his induction coil, to the exploration of their muscles and nerves? We obtain a glimpse of how he overcame that obstacle when he was investigating the effects of stimulating the chorda tympani as it crossed the ear. He succeeded in obtaining opportunities by visiting cliniques where ear cases were treated, and, by holding out the possibility of a cure, he was permitted to apply his electrical method. He thus came by the cases needed for studying the effects produced by Faradisation of the chorda tympani, but to his amazement he found that the patients, particularly in his earlier cases, declared that their hearing was much improved. He thereby opened a field which, had he been a less honest and conscientious observer, might have been, and since his time has been, explored unscrupulously with great financial gain. Experience, however, showed him that such positive results were only obtained in certain cases, and even in them the gain was merely temporary in character. We see in this example how he obtained access to hospital patients and pacified them. The hospital staff-physicians, surgeons, residents, and students presented a still more difficult problem, requiring infinite patience and diplomacy. It was in direct opposition to their experience of human behaviour to believe that a country practitioner should descend on Paris with no ulterior motive, with no aim beyond the mere searching of hospital dust-bins for knowledge. Patience, perseverance, tact, and good-humour ultimately carried the day, and allayed the jealousies aroused by his presence in hospital wards. Awkward situations did arise; questions relating to priority did crop up, and Duchenne was then never slow to defend his own. It was under those circumstances that Duchenne carried out his investigations in Paris from 1842, where he arrived in his thirty-sixth year, until his death in 1875 in his sixty-ninth year.

It does not matter which part or region of the human body we may select to exemplify Duchenne's work; we shall find he has explored it with his rheophores and brought to light a crowd of fresh facts which have a direct bearing on the methods we must adopt for the treatment of its disorders. We might select the muscles of the back which balance the vertebræ on each other and keep the spine erect, or the muscles of the abdomen and thorax, and the part each plays in the act of breathing, or the interplay of facial muscles in expression, but I prefer to choose the region of the shoulder because it was the site of his earlier researches and serves to illustrate the nature of his observations and the practical character of his discoveries. When he stimulated the deltoid muscle, he expected, because his text-books had so taught him, that the arm would be elevated. To his surprise he found it was the scapula and not the arm which moved. He saw that the deltoid could not, and never did, act alone; when it passed into action, messages were at the same time reaching the muscles which fix the scapula. Further inquiry showed him that there was no such thing as a single muscle in the living organism with which physicians and surgeons have to deal; every muscle is woven into a functional complex. The origins and insertions of muscles which so tax the student's application and memory are but the letters of the muscular alphabet; they are meaningless symbols unless we can construe them into words and sentences. When we look at our modern anatomical text-books, can we say we have yet passed beyond the alphabet stage? Hence it was, as Duchenne stimulated each single muscle of the body and watched its individual action, he sought out at the same time its antagonist and associated muscles, for he knew that all of them must move together, answer to the same nerve-reins and be elements of the same neuro-muscular mechanism.

It was in the shoulder, too, he studied the static or tonic action of muscles, and observed that parts of the same muscle, such as the trapezius, may subserve quite different functions. He noted, that when a burden was carried on the shoulder, it was the middle or cervical part of the trapezius which was called into action, while during laboured breathing only the clavicular section took a share in the inspiratory movement. He noted, too, that the respiratory part of the trapezius was much more sensitive to the induced current than the static part. He believed that the poise of the shoulders was a reliable index of the tonic or static action of the muscular system—which, as we have seen, was regarded by Marshall Hall as a condition dependent on the reflex function of the spinal cord. It was Duchenne who first noted that, when we assume the upright posture, muscles have to support and carry the weight of the shoulders and arms; that is their static action. It was he who opened up our knowledge of the peculiar paralysis which follows a lesion of the upper cord of the brachial plexus—the paralysis and wasting of the deltoid, biceps, brachialis, and brachio-radialis muscles. He observed that condition in children whose neck or arms have been strained during the act of parturition. In adults, he noted how the shoulder dropped and the neck became apparently elongated in cases of paralysis of the shoulder muscles and how the head of the humerus fell from the glenoid cavity because the static action of the muscles was destroyed. One cannot read Duchenne's account of the shoulder—as pictured in his great and permanent book, *Physiologie des Mouvements* (1867) —without realising that muscles have no independent or isolated existence; they, their nerves and the central controlling mechanisms of the cord must be regarded as an indivisible whole for him who would seek to help in the repair of the human machine.

"Writers on physiology," said Duchenne, "devote pages to the voice, but of the movements of the hand and foot, which are so essential for daily life, they have nothing to say." He made full amends for the neglect of his predecessors and contemporaries; he tested the action of each muscle of the living hand and foot, and checked his observations by watching the normal movements in the healthy and the defective motions in the maimed. No one in Duchenne's time could give a satisfactory explanation of the position assumed by the fingers in cases of ulnar paralysis—the "claw-hand." Duchenne took up the problem. To guide him he had the anatomical facts with which everyone was familiar. The interosseus muscles of the hand and two, or sometimes three, of the lumbrical muscles were supplied by the ulnar nerve. He had also the general principle on which he worked to serve him as a guide—namely, that in every paralytic deformity not only the disabled muscles had to be studied, but also their antagonists or "moderateurs" and those which worked with them—their "coadjutors." In the clawed hand why was the proximal phalanx of the fourth and fifth fingers hyper-extended and the middle and distal phalanges flexed? "In the first place," he said, "one has to note that the common extensors of the fingers are not the antagonists of the long flexors of the fingers. The common extensor tendon extended only the proximal phalanx of a finger, the flexor sublimis only flexed the middle phalanx and the flexor profundus bent only the nail or distal phalanx." In the clawed hand of ulnar palsy it was plain that these muscles were in uncontrolled action, for the proximal phalanx was extended, the middle and distal flexed. His problem was to discover the antagonist muscles which flexed the proximal and extended the middle and distal phalanges. Stimulation of the interosseus muscles by his Faradic method gave him the solution at once: the proximal phalanx became flexed; the middle and distal became extended. He supposed that the lumbrical muscles had an exactly similar action.[1] "Clawed hand," then, is the result of paralysis of the interosseus and lumbrical muscles. That was one of Duchenne's many discoveries.

It was in palsied hands and arms that Duchenne first noted

[1] In his excellent work on *The Action of Muscles* (1918) Dr. W. Colin Mackenzie states that the lumbrical muscles are the flexors of the proximal phalanges and that the interossei are the extensors of the middle and distal phalanges.

the signs which mark the return of voluntary control. Tonic contraction in the interosseus muscles of the hand was apparent before the patient obtained voluntary control over them. The nearer a muscle was situated to the spinal cord the sooner did signs of recovery appear in it. The first sign of approaching nerve recovery was a hyperæsthesia of the affected part; then the circulation which had become sluggish after section or injury of the nerve began to return, the temperature of the part, which had fallen 5° or 7° below normal, rose to that of the corresponding part of the opposite side; the nutrition of the part improved, the tonic power returned to the muscles, and lastly came, not all at once, but gradually, the power of voluntary contraction. At an early period of his investigations he saw that Faradisation could not replace normal voluntary impulses as a means of keeping a paralysed muscle in health and preventing it from undergoing atrophy; he came to the conclusion that it was only as voluntary control was on the point of returning, that the application of Faradisation could help in recovery. On the other hand, he was of opinion that Faradisation did assist the blood-vessels of a paralysed part in regaining a healthy state.

In 1859, two years after Marshall Hall's death, Duchenne came across his first case of locomotor ataxia. He believed he was the first to recognise this condition as a distinct clinical entity; he was not the first, but there is no doubt that he made the discovery for himself. His previous investigations provided him with the information needed for a correct interpretation of the nature of the lesions in this peculiar disease—a pathological dissection of the functions of the spinal cord. The condition was one clearly connected with a disturbance of the "reflex function," but when he came to interpret the various clinical manifestations it was to Charles Bell and not to Marshall Hall that he acknowledged an indebtedness. Bell's book on the Hand—a Bridgwater treatise published in 1832—contained, he said, one "gem"—the discovery of a muscular sense. Duchenne recognised that in locomotor ataxia this muscular sense was lost; the messages which normally arose in muscles were no longer received to guide and control movements. But, in his opinion, and this is one of his chief discoveries, the "articular sense" was also lost; he presumed that in every act messages or stimuli arose in and around joints—stimuli perfectly distinct from those which arise in muscles—and that it was the loss of these articular reflexes which gave cases of locomotor ataxia their unsteady and peculiar gait. It is not necessary to labour the importance of this discovery for orthopædic surgeons. Hilton, we have seen, based his practice on the assumption that abnormal articular stimuli were the cause of muscular unrest. Marshall Hall's investigations showed why such stimuli were required for reflex movements of limbs; but it was Duchenne's observations in cases of locomotor ataxia that first revealed the extent and importance of joints as signal stations for reflex acts.

The investigation of the shoulder belongs to an early phase of Duchenne's career. We shall now pass to a subject which occupied his later years—the cause and cure of flat-foot, a matter of very grave importance, for every large country loses an army of men from this defect. We must remember, before we enter this later phase of his career, one stretching over the last decade of his active life—from 1862–1872 (for soon after the latter year he was struck down by a hæmorrhage in the brain)—he had passed through a brilliant period of discovery in the previous decade. He had recognised and described as distinct diseases: "Progressive Muscular Atrophy" (which mainly interested him because such cases presented a condition which allowed him to test the action of the deeply situated muscles); Glosso-labio-laryngeal Paralysis; Pseudo-hypertrophic paralysis (for the diagnosis of which he invented an instrument for punching out pieces of the diseased muscles for microscopic examination); and Locomotor Ataxia. There is no need to haggle over his claims to priority; it was no matter to him whether other men had seen and identified such cases before him; he certainly discovered them for himself as he sifted his gleanings in the clinical fields of Paris. The ataxic gait made him think and reason to the very utmost of his power. He saw how much the action of limbs depended on the entirety of their sensory nerves; he observed that the greater the loss of feeling in a limb, then the greater was the inco-ordination of the action of the muscles of that limb. This, he noted, was particularly the case in those patients in whom the articular nerves were affected; he therefore, as we have already noted, looked on the stimuli arising in moving joints as playing a most important part in the regulation of muscular action. It was with such experience behind him and a very intimate knowledge of the action of the muscles of the foot and leg that Duchenne took up the complex problem—one which I regret to say still awaits a solution—the cause and cure of flat-foot.

We can now say with some assurance that Duchenne was the first to show us the right way to study the functional and anatomical collapse of the foot. He studied how its parts worked in the living, standing position. As his own foot swung forward to be planted on the ground in walking, he observed that the muscles which manipulated the foot, used the leg itself as a base of action, but that when the foot became planted on the ground and the weight of the body came to rest on it, the muscles immediately reversed their origins; the foot itself—particularly the great toe—then became the base from which they moved the leg and balanced the body. The balance between the opponent muscle groups had to be perfect if the body were to be poised steadily on the foot. He realised that, as these muscles support and balance the body, they also maintain the arch of the foot; although he does not sufficiently emphasise the fact that the normal arches of the foot are maintained by the action of the pedo-crural musculature, he did most clearly recognise that the breakdown of the arch was due to an exhaustion and loss of co-ordination in the opponent groups of that musculature.

In 1862, a girl of 9, with incipient flat foot, was brought to him. From former observations he had come to the conclusion that the peroneus longus, whose immediate opponent was the tibialis anticus, was the chief agent in maintaining the arch; he regarded the peroneus longus as the "key" of the living arch. On testing the electrical reaction of the muscles of this girl's legs, he found that the response by the peroneus longus in the disordered foot was much less than in the sound foot; the muscle on the disordered side "had become impotent," and was no longer able to balance its antagonist. At a later stage, when pain had appeared in the joints of the foot, particularly in the sub-astragaloid joint, he observed that the peroneus longus was no longer passive, but had entered a contracted or spastic stage, and with this change the flattening of the foot had altered in form. That spastic stage he attributed to reflexes which were being set up in the strained and broken-down joints of the foot. His conclusion was that flat foot was due to exhaustion of the pedo-crural musculature from overwork—from prolonged standing. There was also a "constitutional" defect contributing to the breakdown, but of the exact nature of that defect we are still as ignorant as was Duchenne. The weakness and exhaustion of the peroneus longus being, in his opinion, the primary cause of the collapse of the arch, the first step in treatment must be to strengthen its action by electrical stimulation and to weaken its opponent —the tibialis anticus. He tried to help the weakened peroneus against its opponent by applying—as Barwell and others had done before him—elastic supports to act as an accessory muscle. He found that no elastic band or steel spring could replace a living muscle, no apparatus devised by man could raise or lower its power so as to act in the regulated manner of antagonist muscles. He made one very important and practical observation: he saw that as the arch collapsed, the action of the muscles became vitiated and tended to increase the deformity. Stromeyer had also noted that, but in his opinion the primary cause of the condition was an over-action of the muscles situated in the calf of the leg. The very first step in the treatment of flat-foot—more especially in early cases—was, in Duchenne's opinion, to restore as nearly as can be done the parts of the arch to their normal relationship, so that the muscles which still retain their power may serve, as in the normal foot, as mainstays of the arch. If Duchenne's teaching is right, and in its broad lines I am sure it is right, then those who condemn the use of artificial supports of the arch, applied when its collapse is threatened, do an ill service to their patients. Such a support can give temporary rest to the exhausted muscles; the amount of support may be diminished as the muscles recover.

There is no doubt that Duchenne exaggerated the importance of the part taken by the peroneus longus in maintaining the arch of the foot and in balancing the body upon it. He overlooked the functional importance of the tibialis posticus and flexor longus hallucis because these muscles, buried deeply beneath those of the calf, were not easily accessible to his methods of stimulation and investigation. He was right in supposing that flat-foot is due to a disordered action of the muscles, but time will show, I think, that the chief culprit is not the peroneus longus, but the tibialis posticus.

But of what nature is this muscular disorder? There, too, Duchenne comes to our assistance. It was, in his opinion, of the same nature as the disorder known as writer's cramp—a disorder caused by overuse and exhaustion of certain muscles. The condition of flat-foot is one due to a breakdown in the reflex mechanism which co-ordinates the action—particularly the tonic or static action—of muscular groups which balance the body on the foot. We shall not stop now to inquire at what point of the reflex arc that breakdown occurs—whether at the end organs in which the sensory nerves of muscles commence, or in the spinal centres which control the group; it is enough for our present purpose to show that it was Duchenne who made us look for the cause of flat-foot in a disorder of the neuro-muscular mechanism of the lower extremity, in place of a mere deformation of the bones of the foot.

I might have cited many instances from other parts of the body where Duchenne has added to our knowledge of applied anatomy and physiology. It was he who showed the untenable nature of the hypothesis put forward by the brothers Weber in 1876—that the lower limb acted as a pendulum—as if it were an artificial limb—when it was swung forwards in walking; he demonstrated that the ulna, as well as the radius, took an active and positive part in the movements of pronation and supination; he was the first to observe that "clawing" of the fingers was a result of paralysis of the interosseus muscles, and that as they regained their power of movement the clawing disappeared. Our main object, however, is to see how far he can help us to establish the general principles of treatment—especially how far electrical means may be expected to help us, for he had forty years of experience on which to found his statements. His opinion was that electrical stimuli differed from, and could not replace, volitional stimuli; that paralysed muscles recovered much better under volition than under Faradisation. He was sceptical whether electrical means contributed in any way to the return of voluntary power to muscles, a conclusion which is supported by the modern experiments of Professor Langley; he believed its application was of most service after a muscle had begun to regain movement. He discovered that muscles could be electrically stimulated, not directly, but through their reflex mechanism—by means of electric baths. He was of opinion that, if in some states this "reflex" means of treatment may assist recovery, there were others in which it worked harm. The basis upon which he sought to found his methods of treatment was a knowledge of the healthy mechanisms of the human body; in the prevention and in the rectification of deformities, he sought, as Thomas of Liverpool also did, to assist the weak muscles and restrain the strong.

Here, then, in Duchenne of Boulogne, we have one of the most remarkable figures which has ever appeared on the medical stage. His contemporaries were too close to him to realise that this missionary for science who appeared before them in garb of a rustic country physician, and made his modest bow with a battery of his own design under his arm, was playing a greater part in the drama of Medicine than the star-actors who kept themselves in the centre of the stage and in the full limelight. It was given to him, as is given to few men, to discover a key which could open the door to a new field of knowledge. He used that key, not for his personal aggrandisement, but for the enrichment of medical knowledge. He had what the real investigator needs—patience and perseverance. He had those qualities which the surgeon needs who would undertake the treatment of bodily disabilities and deformity, yet more patience and more perseverance. In the case of the girl suffering from flat foot, which I have already mentioned, he gave sixty sittings before the muscles responded to their normal functions. "Goodness knows," he says in his introduction to the *Physiologie des Mouvements*, "the collection of facts I give in this work have cost me an endless amount of time and pains, and if my life permits, I hope, with the aid of God and the courtesy of my colleagues, to finish what I have put my hand to." It was thus we came by one of the best treatises ever written on the dissection of the living human body.

CHAPTER VII

SOME MATTERS RELATING TO THE MECHANISM AND MANAGEMENT OF MUSCLES

In a previous chapter it has been shown that the discoveries of Marshall Hall in the first half of the nineteenth century made a knowledge of the spinal cord essential for the surgeon who sought to apply rational means in the treatment of orthopædic cases. The reader who wishes to learn how these discoveries should influence orthopædic methods of treatment ought to study the works of Dr. Charles Sherrington, Professor of Physiology in the University of Oxford. In this chapter we shall touch only on the more important facts which seem to have a direct bearing on orthopædic practice. At the end of the chapter the work of a true follower of Duchenne will be touched on—the late Dr. Charles Beevor.

THERE still remain many questions in applied anatomy and physiology which await a rational answer: Why is it, when the patella is broken, that the quadriceps muscle at once shortens and drags the upper fragment away from the lower? The muscles of the calf react by shortening when the tendo Achillis is cut. When a bone is broken, the surrounding muscles contract energetically, so that the broken ends overlap —often a considerable distance. When a joint is dislocated, or when it becomes injured or inflamed, the muscles surrounding that joint pass into a state of hyper-tonus or even of spasmodic contraction. When a group of muscles becomes paralysed, such as those which dorsi-flex the wrist, the opponent group—the palmar-flexors—undergo a contraction or shortening. We must be able to explain the behaviour of muscles under conditions such as have just been cited, if we would adopt rational methods of treatment for their restoration.

On a former occasion [1] I have spoken of muscles as internal combustion engines—the engines which work our limbs and

[1] "Christmas Lectures at the Royal Institution, 1916."

trunk. I have been justly criticised for the use of this simile because it is now ten years ago since Sir Walter Fletcher and Dr. Hopkins discovered that combustion occurs, not when a muscle contracts, but between the waves of contraction, whereas in a petrol-driven engine combustion occurs during the effective stroke of the piston. This difference does not signify that a muscle is not an internal combustion engine; the right interpretation is that a petrol-driven engine and a muscle are *internal* combustion engines of different kinds, for in both, carbon dioxide and water are produced, and the rate of production varies directly with the amount of work done. There is much to be gained, I think, by the use of the engine simile, for to us, who use our muscles in health, they seem simple and easily managed machines; but when disordered, their repair presents us with many a perplexing problem.

Our muscles, if we regard them as living engines, are always running—always ready at any instant to go into motion; their steam is always up. Some of them are running with their clutch out, such as are employed only for temporary or definitely limited efforts; but others of them—the static muscles, those which have to maintain the upright and standing postures of the body—are kept running with their clutch always engaged. The muscles of the back, and of the abdomen, those which maintain the weight of the shoulder and upper limb, those which balance the head, and all the muscles of the lower limb, which maintain the various parts of our body in the standing posture, are "clutched" muscles. But whether they are clutched or unclutched, every muscle must maintain a certain degree of tension in order that it may make an immediate start when called on; that muscular tension must be maintained under all circumstances; there must be no slack to take up before a muscle comes into action. Hence if a tendon is cut or a bone broken, the muscle attached to that tendon or surrounding the fracture must undergo a shortening, for every fibre is under an initial or "starting" tension. The shortening of muscles, in the instances just cited, is a manifestation of the self-regulating mechanism with which all living muscles are endowed.

In this chapter I propose to see how far the researches of modern neurologists help us to understand the mechanisms which regulate the action and behaviour of muscles under surgical conditions. We can do this most easily by confining our attention to a muscle-group which Professor Sherrington explored at an early stage of his career—the extensors and flexors of the knee-joint. All of us have known since the commencement of our medical apprenticeship that a contraction of the extensors—a "knee-jerk"—was elicited when the patellar ligament was tapped, but we, who were students in the "eighties," did not know the mechanism of its production until Professor Sherrington revealed it to us in 1893. It would be an interesting story to tell how he was gradually led up to the investigation of this problem, but we must resist the temptation and confine ourselves to the elements which he found to be concerned in the reflex mechanism of the knee-jerk. Of these the chief are the nerve cells, several thousand in number, which are situated in the anterior horn of the spinal cord opposite the origin of the anterior roots of the 3rd and 4th lumbar nerves. Through the anterior crural nerve, each of these cells sends a process or nerve fibre to establish a connection, by means of its end plates, with certain fibres of the extensor muscles of the knee. To help us to obtain a working picture of the mechanism, we may look on the anterior horn cells as "drivers"—engine-drivers or muscle-drivers—and the anterior horn itself as an establishment set aside for their accommodation. There was a time, no doubt, when each automaton driver was to be found by the side of the particular engine it managed—with its motor plate or hand on the throttle valve of the engine.[1] Although seated now in the spinal cord, their hands are still on the muscular cylinder, and their arms or axons are drawn out into mere threads. I have spoken of the "drivers" as automatons—they are only apparently automatic—they can neither start, accelerate, nor stop their engines, except under the influence of signals or messages, which, we shall see, reach them from many sources.

[1] See Professor Graham Kerr's "Account of the Development of Nerve Fibres in Lepidosiren," *Trans. Roy. Soc. Edin.*, 1906, vol. xli. p. 119.

They also exert a mysterious power over the muscular cylinders immediately under their control; as long as the drivers are in direct touch, their engines have the power of building up energy-yielding material, and, on demand, of liberating energy as work; that power at once ceases when their arms are cut or broken; the engine at once stops and soon begins to undergo a disruption or atrophy by a species of self-consumption. Clearly, in cases where nerves are damaged, our first effort in the restoration of muscles must be the re-establishment of the normal relationship between the anterior horn cell and their group of muscular fibres.

It is clear that the driver cells of the anterior horn being situated so far away from their muscular engines, must be kept informed not only what is taking place in the fibres immediately under their control, but also be supplied with information as to what is happening in all the other associated engines, otherwise they could not handle their machine so as to obtain co-ordinated and efficient results. The manner in which the driver cells of the cord are kept supplied with information from the muscles or engines under their control was discovered by Professor Sherrington.

From his researches we see that each driver cell has attached to it a chain or system of signallers in the shape of nerve cells. Those which subserve the needs of the driver cells, which control the extensor muscles of the knee, are situated in the posterior root ganglia of the 3rd and 4th lumbar nerves of the human body. Each signaller cell in a posterior ganglion is connected by one process or wire with a receiving station or receptor organ, technically known as a "muscle-spindle," situated in a particular group of extensor fibres of the quadriceps. By its other process or arm, these signaller cells of the posterior root ganglia transmit their messages to various stations in the spinal cord. In the vastus internus muscle, particularly under the fascial expansion which couples it to the knee-cap, Sherrington found muscle spindles or signal stations to be particularly numerous. He observed that they were so placed that they served to register in the most sensitive manner every change in tension which may occur when the vastus fibres are in action. Every change thus registered at every one of the many receiving stations is transmitted by signallers in the posterior root ganglion to the spinal cord—ultimately reaching the driver cells in the anterior horn. Through such messages driver cells are kept informed of the working state of their own engines.

Messages are transmitted from the posterior root signallers to the driver cells in the following manner: Between the posterior root signaller and anterior horn driver are placed signallers belonging to a different category—to an "intermediate" or "exchange" group. Each "exchange" signaller evidently taps the messages which arrive in the spinal cord, selecting such as are destined for their particular clients amongst the driver cells. Thus each motor or anterior horn cell is made aware of what is happening in its own particular series of contractile cylinders by means of a chain of signallers. The signallers and motor cells are so arranged—so linked hand to hand—as to form a loop or arc—a reflex arc. A muscle can thus be started by a stimulus or message delivered in receiving stations situated amongst its own fibres, and once started, its further action may be regulated by further messages automatically set up by the contracting fibres and transmitted to the driver cells. Such messages or stimuli are duly transmitted to the appropriate driver cells by signaller cells.

There are also other special signal-receiving stations (Golgi bodies) placed in tendons. We have thus a machinery which, set into action, gives us a rational explanation of the clinical phenomenon known as a knee-jerk. When the patellar ligament is tapped, the tension of the muscle spindles within the extensor muscles of the knee is suddenly increased; that increase is signalled to the anterior horn cells and evokes there a message or stimulus which throws the muscle into a state of action. The knee-jerk is an artificial kind of movement—one which we elicit by playing upon the spinal or reflex controlling mechanism of the extensor muscles of the knee. The important fact we learn from Sherrington is that in every contracting muscle messages are being evoked and transmitted to the spinal cord, where they are duly conducted to the "driver" cells of the anterior horn. Every time we rub, pinch, or handle a muscle we excite streams of those "driver-controlling" messages. Every stroke given by the hand of the masseuse is accompanied by messages streaming to the cord, and reflected from there back to the part in which they were aroused.

When the extensors of the knee contract, messages pass from their muscle spindles, not only to the quadriceps "driver" cells in the anterior horn of the spinal cord, but also to the cells which control the opponent group of muscles, the hamstring flexors. The reason for this is very obvious. Muscular engines unlike any form we have evolved for industrial life, are set so as to work in opposing groups. It is quite clear, if there is to be no loss of power, that the work of opposing groups must be co-ordinated with the utmost nicety. Hence the need of messages passing from the muscle spindles of the extensors of the knee to influence the driver cells which preside over and regulate the contraction of the hamstring flexors of the knee. Sherrington was able to prove that a constant stream of messages arises in the contracting extensors and, passing to the cells which control the action of the hamstrings, exercises on them a regulating influence. In this way a balance is maintained between the degree of shortening of the extensors of the knee, and the degree of lengthening of the hamstring flexors. Similarly, messages are being produced in the hamstring muscles which exercise a regulating or inhibitive influence over the extensor motor cells. Thus the state or degree of contraction in any muscle is regulated in the main by two sets of stimuli or messages: (1) Those arising in its own fibres; (2) those arising in the fibres of its opponent. The more nearly we can place muscles in a state of rest, the less call will there be on this reflex mechanism.

Professor Sherrington observed, when the hamstrings were relaxed, or when their afferent (signaller) fibres had been cut, that the action of the quadriceps extensor, as elicited by the knee-jerk, was increased. The jerk was increased because the stream of inhibitive messages from the hamstring muscles had been cut off. On the other hand, when the hamstrings were in a state of contraction, or if their afferent nerve fibres—going to the signaller cells in the posterior root—were artificially stimulated, the extensor response, as manifested by the knee-jerk, was greatly restrained. The restraint was due to the increased stream of inhibitive messages passing from the hamstrings to exert their influence on the "driver" cells which control the extensor muscles of the knee. We thus learn that every time the hamstring muscles are kneaded or massaged, stimuli are evoked in them and transmitted by the signaller cells of the posterior root ganglia and of the spinal cord to their controlling or driving cells of the anterior horn. Corresponding messages are also forwarded to the driver cells of the opposing group of muscles—the extensors of the knee. Stimuli, excited in the manner just mentioned, tend to set up a state of contraction in the hamstring group and a state of relaxation in their opponents—the extensors of the knee. If a state of flexion of the knee-joint be already present, the application of massage will increase the degree of flexion, not lessen it. We see, too, that once a muscle is thrown into action, its automatic signal mechanism tends to keep it in action.

It is becoming sufficiently apparent that those muscular engines which are so often damaged in military as in industrial warfare, and so often tax the surgeon's care, are very complex machines. The arrangement and function of the serried battalions of driver cells situated in the anterior horn of the spinal cord, seem simple enough. The battalion of driver cells which controls the extensor muscles of the knee—the battalion on which I want to centre the reader's attention—is placed opposite the origin of the anterior roots of the 3rd and 4th lumbar nerves from the cord. Into the battalion quarters of these driver cells are pouring the streams of messages which arise in the receiving or sensory stations of their own muscle—the quadriceps extensor cruris. Into the same quarters are coming messages which arise in the antagonist group—the hamstring—and also by a less direct route, numberless messages which are excited in more distant muscles—those which co-operate with the extensors of the knee—in the various movements needed in standing and walking.

For the surgeon, the messages which arise at the articular receiving stations are of the first degree of importance. Round the knee-joint, particularly in its ligaments and capsule, is placed a series of end organs or receiving stations, furnished with the onion-like wrappings of a Pacinian corpuscle. They are so placed and shaped as to be affected by all changes in pressure and tension produced by movements of the joint. The messages arising in the articular signal stations of the knee, bear in upon the extensor driver cells, and keep them in touch with what is happening in the joint with which they are most directly concerned—the knee-joint. We can see how these articular signal stations of the knee may give rise to false or pathological messages when ligaments become strained by the locking of an inter-articular cartilage, or altered by inflammation—messages which throw the driver cells into a state of disordered action. When the knee-joint becomes inflamed, particularly when the articular cartilage is eaten away and the granulations springing from underlying bone are exposed, numerous new signal stations of a pathological nature are established—stations at which painful messages are freely excited. Such messages have apparently a particularly easy access to the driver cells of both the extensors and flexors of the thigh, setting up an inco-ordinated struggle in which the hamstrings win because they have the more powerful leverage. Hilton knew well, if he could arrest the production of painful articular stimuli within the knee-joint, which he sought to do by keeping the joint at rest, that he would succeed in allaying the contracted state of the muscles of the thigh.

I have not yet finished with the signal stations from which the automaton drivers of the muscles of the knee are furnished with guidance. If the skin of the foot or leg be pricked, the flexors of the thigh and knee are thrown into action. They may also be directly influenced by cutaneous impressions arising in the region of the knee. The skin over the knee is provided with several different kinds of nerve endings, or stations, which, when suitably stimulated, give rise to signals of touch, pain, heat, cold, etc. Such cutaneous messages may also reach and affect the anterior horn cells of the knee region of the spinal cord. Hilton proposed, in cases where cutaneous nerve stations in the region of the knee were hyper-excited, to arrest or diminish the transmission of stimuli by the local application of belladonna or opium in strong solutions. It is also clear that these cutaneous signal stations and their attached signallers can be hyper-excited by the application of counter-irritants to the skin of the knee, and that an effect of an uncertain nature can be produced on the driver cells of the cord, and, through them, on those engine cylinders which rest under the control of such cells.

We cannot overstate the importance of this complex silent mechanism, which regulates the movements of our limbs, because it is by understanding it and using it that the orthopædic surgeon hopes to find a rational system of treatment. Why should such a mechanism be provided for the human body? Think for a minute what the conditions of life would be if every moment we stood erect, or every time we took a step, we had consciously, and by an effort of will, to set into motion, and regulate at every phase of that motion, the millions of muscular fibres which the act of standing or walking calls into action. If our limbs were so controlled life would be intolerable, and movement only possible for a genius. Hence our muscles have been transferred to an automatic mechanism over which our wills have the right and power to intervene at any moment. In health that automaton is a blessing, and in disease it is of distinct advantage for the surgeon, because it is by understanding and utilising the natural mechanism of muscles that he hopes to attain the condition of rest necessary for a return of health.

In an earlier part of this chapter mention was made of the intermediate cells of the spinal cord—the "exchange signallers" —which linked up the signaller cells of the posterior root—"posterior root" and the driver cells of the anterior horn. These three elements make up the reflex spinal arcs which control all muscular acts of trunk and limb connected with the primary needs of the animal body. Sherrington's "spinal dog" can carry out elaborate purposive movements, and yet the underlying mechanism which carries out these movements is merely the reflex-arc system of the spinal cord. The most remarkable element in the arc system is the "intermediate" or "exchange" signaller cell of the spinal cord. A cell of this kind seems to possess much the same kind of powers as the intelligent operator of a central telephone exchange. An intermediate signaller cell has particular clients amongst the driver cells of the anterior horn: its proper duty is to supply them with messages coming in from particular receiving stations in the body—an area of muscle, joint, or skin, with which that particular cell has to do direct business. Such is the routine occupation of all exchange signallers. They also possess another function, equally important. They are the recipients of messages coming from the brain (volitional), from the cerebellum, from the labyrinth, from the corpora striata and tectate nuclei. The condition of a muscle, or of a group of muscles, at any particular minute, is the result of the messages which the intermediate cell is gathering, receiving, and forwarding to the driver cells which control that muscle. Duchenne showed that the individual muscle had no real existence in the living animal body; its action is fused with that of a crowd of coadjutors and antagonists, which by their combined and balanced action execute a defined and purposive movement. Marshall Hall and Sherrington have convinced us that, in a functional sense, we cannot separate a muscle from the neural-arc system; the neural arc and muscle are parts of a single apparatus. Both make up parts of one organ. That is a truth which has to be kept in mind, both in the diagnosis and in the treatment of injured limbs. If the arc system becomes damaged at any point, there is an immediate breakdown in the muscles connected with that system—the nature of the disturbance depending on the exact site of the lesion.

The reflex system of nerve arcs just described forms the machinery by which the surgeon hopes to restore normal movements to every muscle which retains the merest flicker of a contraction, one so slight that the finger may succeed in detecting, yet the eye fail to see. In our efforts to restore normal action to a damaged limb we base our hope very largely on the adaptative powers—amounting almost to an intelligence—possessed by the exchange signallers of the spinal cord. If we can get a message through to them from the muscle, they will, given time and a few remaining intact driver cells, open and extend a motor path to the muscle. For many reasons the contraction set up in a weakened muscle by a volitional stimulus is better than that set up by a faradic or galvanic stimulus—because only volitional impulses can set groups of muscles into normal action—evoking in the prime movers the normal stimuli which regulate the elongation of the antagonistic group.

We have to remember that a voluntary impulse not only plays on the muscle-controlling machinery of the spinal cord, but also on another reflex mechanism—the vaso-motor. The extension of the two systems—motor and vaso-motor—to any muscles must work in absolute harmony. The greater the work a muscle is called on to do, the more must its supply of blood be. We know that a volitional impulse which sets the muscle working, also sets the mechanism which regulates the blood supply going at its proper rate; but we have no evidence that this double effect can be accomplished by any artificial form of stimulus, electrical or other.

What are the best means of maintaining in health a muscle which has just become paralysed—particularly one to which we may expect an early restoration of nerve supply? Let us look at the condition of such a muscle: the influence of the anterior horn and posterior root cells has been withdrawn from every muscular fibre; the vital contents of every cylinder begin to break up, degenerate, and be absorbed. The influence of the vaso-motor system is also lost; the capillaries become unregulated, irregular, distended channels. We wish, as far as possible, to prevent the breaking up of the muscle cylinders and to maintain intact their original condition so far as we may. Massage applied to the delicate and broken organisation of a crippled muscle seems to me calculated to help in the work of destruction. If in any crisis of surgical treatment rest is necessary, it is in the early stages of degeneration of a paralysed muscle.[1] Surgeons are misled by the fact that the paralysed muscle does not complain when it is kneaded and rubbed. How could it? Its nerves are cut. I am certain that Dr. Colin Mackenzie is right when he applies splints as carefully to a paralysed muscle as to a broken bone; perhaps the muscle is in the greater need of this protective treament.

Why is it that a paralysed limb—or a paralysed muscle—is so sensitive to cold and needs warmth as well as rest? The causes are multiple; muscles which are healthy and capable of active contraction are the best heat producers and heat radiators of the body. The greater the muscular paralysis of a limb, the greater its susceptibility to cold. The other cause which is at work is that division of the principal nerves of a limb destroys the afferent and efferent paths of the vaso-motor mechanism of the limb—a mechanism which is employed to maintain an equable temperature in a healthy limb. Such a paralysed limb has lost its self-regulating heat mechanism; its temperature rises or falls, like that of a lizard, with the surrounding atmosphere. A paralysed muscle has lost its heat-regulating mechanism; it stands as much in need of warmth as does a newly born babe.

Another bearing of the vaso-motor mechanism on the work of the surgeons is its possible connection with the establishment of a collateral circulation. Why is it, when we tie the femoral artery in Hunter's canal, that within a week minute arterial channels, which link proximal with distal branches of that vessel, begin to open up? The pressure of blood within the patent part of the femoral artery has not changed. It is possible that the vaso-motor mechanism of the limb comes into action, but that alone would not suffice to give us the new channels we observe developing. We have to assume that the starved tissues beyond the site of ligature have the power by some undiscovered mechanism, of so acting on the

[1] For further evidence see Dr. W. Colin Mackenzie's *Action of Muscles*, 1918.

available blood channels, that they continue to grow in calibre until a sufficient blood supply is given to them.

We are still ignorant of the exact arrangement and distribution of the vaso-constrictor and vaso-dilator nerve fibres of the limb. A number of years ago Professor Wingate Todd's attention was drawn to this matter by one of those cases of cervical rib which was attended by a vaso-motor disturbance and by pain, due to pressure of the lowest brachial trunk against the abnormal rib. He observed that a strand of fibres from the sympathetic system joins the lowest trunk of the brachial plexus, and inferred that the pain and vascular symptoms were probably due to pressure on these vaso-motor fibres. A fuller account of observations by him and by his pupils will be found in the *Anatomical Record*, May 1914, and the *Lancet*, 10th August 1912. The median nerve in the upper extremity and the posterior tibial in the lower, are laden with vaso-motor fibres (unmyelinated) destined to regulate the blood supply to the palm and fingers in the one case, and to the sole and toes in the other. The fibres in the median and posterior tibial nerves are apparently the efferent elements of the vaso-motor arc; the stimuli which they carry cause dilation or constriction of the vessels. But which are the fibres which serve as the afferent paths to the vaso-motor reflex system? They are undoubtedly those unmyelinated fibres which commence at cutaneous receiving stations or end-organs of various kinds—those at which "painful," "hot," or "cold" stimuli are received.[1] Everyone is familiar with the fact that pain is very closely connected with vaso-motor disturbances. Vascular dilation attended by extreme pain may be produced in the hands or feet by the application of mechanical violence or extremes of heat or cold. Such results are obtained through the vaso-motor reflex system. Painful stimuli may produce a localised vaso-dilatation, that localised vaso-dilatation, if within tissues which are confined by rigid or semi-rigid walls, will still further favour the production of painful stimuli, and so a vicious pain-producing reflex circle is

[1] See researches by Drs. Ranson and Billingsley, in the *American Journal of Physiology*, 1914, vol. xxxviii. p. 128; 1916, vol. xl. p. 571.

produced. Hilton produced pain in an ulcer of the leg when he compressed the vein leading from the ulcer; the pain was diminished when the artery leading to it was compressed.

Dr. John S. B. Stopford[1] has drawn the attention of military surgeons to a burning, persistent and intolerable pain —thermalgia—which occasionally occurs in the hands or feet of soldiers. The pain is usually associated with a partial wound of the median (if the hand is the seat of pain) or in the sciatic or posterior tibial in the case of the foot. Dr. Stopford regards the pain in such cases as being reflex in origin, manifested through the vaso-motor arc system of the limb. Pathological stimuli are produced at the wound of the nerve trunk by cicatricial contraction. The stimuli act on the afferent fibres (unmyelinated fibres of the protopathic system) which excite the vaso-dilator fibres of the foot or hand. One also sees similar reflex disturbance of the musculo-motor system of a limb. A soldier may sustain a wound of his median nerve so that all the muscles supplied by that nerve in the hand and forearm are paralysed. On the other hand, the neighbouring muscles supplied by the ulnar nerve may be in a state of over-contraction. The ulnar contraction in such cases is apparently due to artificial or pathological stimuli arising at the wound of the median nerve, passing into the spinal cord and there stimulating, in a reflex manner, the anterior horn cells from which the motor fibres of the ulnar nerve arise.

Having thus made an imperfect survey of the modern development of the line of research opened up by Marshall Hall in 1832, so far as that research bears on orthopædic treatment, we shall now turn for a moment to note what has happened in the field of inquiry opened by Duchenne in the year 1835. Duchenne made the movements of the human body his special study. He was succeeded by E. J. Marey, who perfected a new way of studying muscular movements—first by taking graphic records of them by various forms of registering machines (La Méthode Graphique, 1878), and, later, in 1883 by photographic records—the cinematic method. When

[1] *Lancet*, 1917, ii. 195.

we survey our British field of medical inquiry we find only one man who has followed up this line of research—the late Dr. Charles E. Beevor.[1]

We must note the circumstances which led Dr. Beevor to follow in the track marked out by Duchenne. Like so many of our leading British neurologists he was a student of University College, London, and a physician to the National Hospital for the Paralysed and Epileptic, Queen's Square, Bloomsbury. In 1878, when he was house-physician to Dr. Bastian at University College Hospital, he had as one of his clinical clerks the late Sir Victor Horsley—the beginning of an association which proved important for both men. Some six years later, when Beevor was 30 years of age and recently appointed physician to the hospital in Queen's Square, he and Horsley made together a prolonged experimental inquiry with a view of determining the exact localisation of movements in the cortical areas of the monkey's brain. It was during this inquiry that Beevor's attention was directed to the importance of noting the various muscles concerned in each movement evoked by stimulation of the cerebral cortex. The result of that inquiry was to show him that Duchenne was right; a single muscle has no existence in the physiology of the body. It also convinced him that Hughlings Jackson was right when he taught that the brain knew nothing of muscles, only of movements. Thus we find that Beevor approached the study of muscles through the central nervous system, whereas Duchenne approached the central nervous system through his observations on muscles. Beevor began by stimulating the cortical areas connected with muscular movement by the Faradic current, but when he came to study muscles he relied, as we shall see, on purely clinical methods. Duchenne applied the Faradic current to the study of muscles, but in his investigations into the central nervous system he relied chiefly on clinical methods.

The value of Dr. Beevor's work[2] to the orthopædic surgeon

[1] Dr. Beevor died 5th December 1908, at the age of 54.

[2] See his Croonian Lectures, *Brit. Med. Journ.*, 1903, i. pp. 1357, 1417, 1481.

lies in his method of studying both the normal and disordered action of muscles. We can exemplify his methods best by selecting some of his shoulder cases. A young girl is brought to him with the trapezius on the right side atrophied below the level of the spine of the scapula. When she began to advance her arm, raising it up until it became horizontal, he observed that, as soon as the movement began the lower angle became prominent, the degree of winging increasing until the arm was half-way up, at an angle of 45°, and then the winging disappeared. "The explanation I would offer for this occurrence," he writes, "is that, though the serratus magnus is mechanically in the position to fix the scapula and prevent its lower angle being rotated towards the spine, it is not its function when the movement is one of advancing the shoulder to act on the scapula, until the humerus has been moved by the deltoid through about 45°. The lowest part of the trapezius is the proper muscle to fix the scapula in advancing the humerus, and when these inferior fibres are paralysed, the serratus will not slip into the breach, so to say, and do the work for the trapezius, which, consequently, is not done at all." No amount of anatomical reasoning or of electrical stimulation could tell us the respective functions of the trapezius and serratus in moving the scapula; their particular functions can only be learned, as Beevor learned them, by direct observation on the living. On the other hand, when the shoulder joint is anchylosed, the serratus magnus begins to act as soon as the arm begins to be raised.

He discovered a physiological method of throwing a muscle from a state of contraction into one of instant relaxation—a discovery which he found useful for testing the degree of action possessed by a muscle which is suspected to be partially or completely paralysed. When the forearm is bent at the elbow the biceps is contracting and shortening, while its antagonist—the outer and inner heads of the triceps—is in a state of contraction but undergoing elongation. Beevor observed that if a resistance is offered to the bending of the forearm, the biceps, as before, goes on contracting, but, as may be detected by the fingers placed behind the humerus, the triceps passes at once into a state of relaxation. Over and over again he found that the "resistance test" helped him to detect defective action in muscles. To test the deltoid, for example, he raised the patient's arm away from the body until it was almost horizontal. Then placing the fingers of one hand on the muscle and holding the patient's arm with the other so as to offer a resistance to its descent, he asked the patient to make an effort to reappose the arm to the side. The moment that effort was made the antagonists of the deltoid were brought into play, and the deltoid, if it possessed the power of contraction, would be felt to relax under the physician's fingers.

Beevor made a particularly close study of the muscles which act on the wrist and on the fingers. The researches of Hunter and of Duchenne revealed the complex nature of the action of the long flexors and extensors of the fingers. When the flexors contract, so as to bend the fingers in grasping, the extensors of the wrist also come into action, to fix the carpo-metacarpal base on which the fingers are poised. Similarly, when the long extensors contract, to extend the proximal phalanges, the flexors of the wrist pass simultaneously into action. Beevor was struck by the fact that when the fingers are bent and the long extensors are brought into action, the extensors of the carpus also contract, but if the fingers are unbent, and then the long extensors of the fingers brought into action, the extensors of the carpus no longer give any assistance. He found, however, he could bring them into action by offering a resistance to the bending upwards of the fingers; when the fingers felt themselves overloaded, they had a means of calling the extensors of the wrist to their aid. As the extensors of the wrist came into action, Beevor could feel their opponents, the flexors of the wrist, pass out of action. By this and many other observations, Beevor helped to reveal the very complex mechanism we have to understand and restore in the treatment of disabled joints.

I will select only one other example to show how well Beevor's observations deserve our close attention. He brought home to medical men that in the movements of the trunk the force of gravity was almost as important as the action of muscles. Indeed, in most movements of the trunk—in bending forwards or backwards, or on flexing the body sidewards—the true moving force was gravity, and the muscles of the trunk served merely as active antagonists. When we bend forwards it is not the rectus abdominis which passes into contraction, but the muscles of the back; when we bend our bodies towards the left it is the muscles of the right side which pass into action. The action of the muscles of the trunk is that of antagonists—they contract, and elongate as they contract. Beevor showed, from his studies of hemiplegia, that the muscles of the right half of the trunk are thrown into action by the *right* half of the brain when the body leans over to the left, but when they begin to raise the body to bring it back to the vertical again, then these same muscles pass under the domination of the opposite hemisphere—the left. In all of those observations there lies a basis of hope for the physician and surgeon. It is not muscles, but movements, which are represented in the brain. The brain, by dint of repeated effort on *the part of the patient*, can be educated. The future will reveal how far education may restore paralysis of central origin.

CHAPTER VIII

THE DEGENERATION AND REGENERATION OF NERVES

The treatment which the surgeon of to-day applies for the recovery of injured nerves is based on the observations and discoveries of a great number of men. The chief events are connected with the lives of Joseph Swan (1791–1874), Sir James Paget (1814–1900), Marie Jean Pierre Flourens (1794–1867), Augustus Waller (1816–1870), Francis Maitland Balfour (1851–1882), Wilhelm His (1831–1904), Jean Joseph Emile Létiévant (1830–1902), and Dr. Henry Head. In the present chapter the reader is introduced to the part which each of these men played in the development of our knowledge of Nerve Degeneration and of Nerve Regeneration.

OF the thousands of maimed men who flood the orthopædic hospitals of Europe at the present time, none tax the knowledge, ingenuity, and patience of the surgeon so much as those who have lost the use of a limb from injury to a main nerve trunk. A nerve divided by a bullet which drills a *clean* track through the soft parts, needs time for its recovery, but if the limb be kept warm and the muscles which are supplied by the injured nerve be protected from the over-action of their healthy antagonists, we can rest assured of a successful result. Unfortunately, such cases are rare; the missile is only too frequently a fragment of shell which splinters and pulverises the bone, tears the nerve trunks, lacerates and crushes the muscles, leaving behind it a churned-up and infected mass of damaged tissues in which sepsis runs riot; and if the surgeon succeeds in saving the limb it is one in which the nerve trunk at its point of rupture is embedded in a mass of cicatricial tissue; the muscles are atrophied, damaged, and bound down by adhesions; tendons are fixed to their sheaths, and perhaps the neighbouring joints have also become involved and stiffened. My aim in this chapter is to examine our stock of knowledge relating to the repair of nerves, and to ascertain which part of it is likely to be of practical use in such cases.

In 1819—almost a century ago—the Council of the Royal College of Surgeons awarded the Jacksonian Prize to Joseph Swan for an essay on "The Treatment of Local Morbid Affections of the Nerves." At that time Joseph Swan was 28 years of age, and for five years had been assisting his father in Lincoln, and was expected to carry on a line which had supplied Lincoln with medical men for several preceding generations. Fate, however, had willed otherwise, for when a student at the twin hospitals in the Borough—St. Thomas's and Guy's—he had fallen under the spell of anatomy, and that spell —as it turned out—stood ever afterwards between him and success in practice. When he was 40 years of age he came to London and fixed his door-plate to a house in Tavistock Square, turning a spacious billiard room into a private dissecting room. There he was allowed to carry on his investigations on the Nervous System in peace, for few patients came to consult him, and the great hospitals of London could find no vacancy on their staff for him.

His later work [1] is of no importance to us now, but it is otherwise with a single chapter in his early prize essay of 1819. This chapter is entitled "An Experimental Enquiry into the Process Nature employs for repairing Wounds of Nerves." It is a chapter in the history of surgery, for these experimental sections of the sciatic nerves, which he performed and observed in twenty-two rabbits, are the first records of their kind. He tells us his reasons for making these experiments. His own opinion was, and it was the prevalent opinion of his time, that when a nerve was cut its ends became reunited, and when the union was complete the nerve could again serve as a highway for messages to and from the brain. He thought the experiments which Abbé

[1] Later he retired from London and died at Filey, Yorkshire, in 1874, aged 83. He was the senior fellow of the Royal College of Surgeons at the time of his death, having become a member in 1813. His essay, first published in 1820, was republished in 1834 as *A Treatise on the Diseases and Injuries of Nerves.*

Fontana and Cruikshank had carried out in London, some of them in John Hunter's house in Jermyn Street,[1] had definitely settled the matter. Jacques Delpech of Montpellier thought otherwise; it was true, he said, the ends of cut nerve became united by a cicatrix, but across that cicatricial bridge messages could not pass, and hence a cut nerve was a nerve permanently paralysed. As a country practitioner in Lincoln, Joseph Swan had experience of horses; he knew that if a horse went lame in a foot from "side-bone" then section of the nerves at the pastern arrested the pain and therefore removed the cause of the lameness. He observed that in from six to eight weeks the foot again became sensitive and unsound, from which he inferred that in that space of time the nerves had healed and could again transmit messages. He knew that veterinary surgeons, in such cases, cut out an inch from each nerve, but even when such gaps were made he had seen lameness to recur at the end of six months. Thus, before he commenced his experiments he had good reason to think that a divided nerve could and did recover its normal function. He resolved, however, to take John Hunter's advice—"not to think but to try." He first studied the actual process of healing, exposing and cutting the sciatic nerve in the middle of the thigh, and killing the rabbits at stated dates after the operation to note what happened at the site of the lesion. He found that the cut ends of the nerve became red, vascular and swollen—particularly the upper or proximal end; the breach between the ends was filled up by an inflammatory exudate—"coagulable lymph"—into which the vascular network on the divided ends spread until they fused and thus the gap in the sciatic nerve was bridged by a span of newly organised vascular tissue. The bridging tissue became firmer and denser, and at the end of a period of about eight weeks he noted that the movements at the ankle and foot began to return, and he therefore inferred that the new bridge had become fit for the passage of nerve messages from and to the brain. But he also observed that

[1] See Chapter I. for Cruikshank's paper. See *Phil. Trans.*, 1795, vol. xxxv. p. 117. The paper was read at the Royal Society, June 13, 1776.

even at the end of eighteen weeks movements were not perfect. He noted, too, that if the sciatic nerve were punctured, or if only a partial division was made, that he could detect no imperfection in the movements of the foot or ankle, and that such wounds healed up exactly as if a tendon had been incised. If, however, he removed half an inch of the sciatic nerve, although the gap did become filled up, he observed that restoration at the best was only partial or else completely failed. In one rabbit, from which he had excised half an inch of the sciatic nerve, he obtained a result of particular interest and importance to us. When he killed the animal at the end of four months, it had an ulcerated heel and "was certainly much improved in the use of the limb, but it was far from being perfect"—he found a gap of two-thirds of an inch (16 mm.) between the cut ends, but this gap was bridged by two nerve filaments—one going to the internal popliteal (tibial) division, the other to the external popliteal (peroneal) division. He also noted that the nerve to the outer side of the heel—the communicans peronei—which came off from the sciatic just above the point of section, was greatly enlarged. Thus we find, four years after Waterloo was fought, and when there must have been scores of men in England with wounded nerves, a young provincial practitioner offering his profession, as a gift, a rational basis for the healing and treatment of nerves. Yet, as the sequel will show, that gift, if accepted, was never turned to any practical account until nearly a century had passed. The seed which Joseph Swan sowed failed to sprout.

We now pass on to an event which took place in the theatre of the Royal College of Surgeons—one at which Joseph Swan was probably present, for he was a member of its Council at the time when James Paget, an East Anglian, with a thoughtful face, a piercing eye, sharply chiselled features, and the slight stoop of a student, entered to give the first of a wonderful series of lectures. He had made himself master of the English tongue, so that his thoughts fell upon his hearers in a clear, consecutive, and finished and particularly pleasing sequence. At the point we are making his acquaintance, Paget was 33 years of age, a young man as surgeons go; he had just been appointed assistant-surgeon to his hospital, St. Bartholomew's, and had performed his first operation in private practice. We have met him before, at the beginning of the tenotomy movement in 1839, when he was carrying out experiments to discover the manner in which union was effected between the ends of divided tendons. At that time he had already fixed his attention on the great and growing physiological school which Johannes Müller had built up in Berlin. He was quick to see the revolution which the "cell-doctrine" and the microscope were to effect in all subjects pertaining to the art of healing. When, in 1842, his hospital created a lectureship in Physiology and appointed him to fill it, he made himself master of the best that the laboratories of Europe had to teach. In the same year (1842) he was selected—a selection as fortunate for the Royal College of Surgeons as for himself—to prepare a catalogue of the pathological specimens in the museum—including Hunter's great collection. That task, which occupied his spare hours for seven years, brought him into intimate touch with the best work of a previous generation. Thus, when he appeared in the theatre of that College in 1847 he was already master of the best that the past and present could give him.

We turn at once to the lecture in which he describes the manner in which a divided nerve is healed.[1] "Healing of divided nerves," he said, "may be accomplished in two methods, which may be named, respectively, primary and secondary union. . . . The secondary healing of divided nerves presents many features similar to that of divided tendons. A bond of new substance is formed, which connects the ends of the retracted portions of the nerve, and in which, though at first it is like common reparative material, new nerve fibres form, and connect themselves with the fibres in the portions above and below. Twelve months generally elapse before, if ever, any restoration of the function is observed." Thus we find that it was Paget's opinion that nerve fibres could arise in the reparative material which filled the

[1] *Lectures on Surgical Pathology*, 1st ed. 1853.

gap in a divided nerve, and that such fibres could effect a union with the original fibres situated on each side of the gap.

In primary union of a divided nerve he believed that a totally different process took place. If the divided ends were in contact, then immediate union might occur between the nerve fibres thus brought into apposition, so that "the nerve could conduct in a fortnight, and perhaps much less, after the wound." His reason for believing in the possibility of immediate union of nerves was founded on evidence derived from two remarkable cases. "A boy, 11 years old, was admitted into St. Bartholomew's Hospital, under Mr. Stanley, with a wound across the wrist. Half an inch of the upper portion of the divided median nerve lay exposed in the wound, and was distinctly observed and touched by Mr. Stanley, myself, and others. All sensation in the part supplied from the radial and median nerves below the wound was completely lost directly, and for some days after, the injury. The edges of the wounded integuments were brought together and in ten days or a fortnight the boy began to observe signs of returning sensation in the parts supplied by the median nerve, and these increasing, I found, a month after the wound, that the nerve had nearly recovered its conducting power. When he was blindfolded he could distinctly discern the contact of the point of a pencil with his second finger and the radial side of his third finger; he was less sure when his thumb or his fore-finger was touched; there were a few and distinct small portions of the skin supplied by the median nerve from which he still derived no sensation at all."

The other case related by Paget to prove the primary or immediate union of a divided nerve is even more remarkable. "A lad, near Market Harborough, 13 years old, had his hand nearly cut off at the wrist joint by the knife of a chaff-cutting machine; the hand was attached to the forearm by only a portion of integument about an inch wide, connected with which were the ulnar vessels and nerve, and the flexor carpi ulnaris muscle—all uninjured. The hand and arm were brought into apposition and retained firmly with adhesive plaster and a splint of pasteboard. The warmth of the hand returned; in ten or twelve days after the injury there was a slight sensation in the fingers, but in the thumb none was discernible till more than a fortnight had elapsed. Finally, the sensation of the hand and fingers, and most of their movements, were perfectly restored."

"Now all this proves, said the lecturer, "that the ends of the divided median nerve had coalesced by immediate union, or by primary adhesion with only an exceedingly small amount of a new substance formed between them." We shall find as we proceed that throughout the second half of the nineteenth century cases of divided median nerve, manifesting the same signs as those narrated by Paget, were reported by surgeons and quoted as evidence of immediate union of nerves. It was not until the opening of the present century that the signs of early return of sensation found a full and satisfactory explanation—not in an immediate reunion of nerve elements, but by the recognition of a system of nerve fibres which Sir James Paget and his successors had overlooked.

Although Paget's theory of immediate union has turned out to be wrong, it led him to suggest a procedure which has proved of real service—namely, the suture of nerves.. As a matter of fact, in 1828—almost twenty years before Paget commenced his lectures at the College of Surgeons—the practice had been applied, not by a hospital surgeon, but by a brilliant experimental physiologist, the handsome, courteous, clever and conceited Marie Jean Pierre Flourens. He was one of those gifted, precocious youths who are born to open up and explore new fields of knowledge as easily as most of us open up a limited field of livelihood. Becoming a Doctor of Medicine at Montpellier in 1813 when he was 19 years of age, he hastened to Paris, threw Medicine to the winds, and joined the Academical group of men who centred round Cuvier, Lamarck and the elder Geoffroy St. Hilaire. The field he chose to explore was the obscure and difficult one relating to the physiology of Sensation. By deftly planned experiments he demonstrated that, with the removal of the cerebrum, all initiative was lost; that the cerebellum was concerned in the co-ordination of the muscles; that the semicircular canals were for balancing; that there was a point—a "nœud vital" —in the medulla oblongata which, when pricked, caused death; that stimulation of the spinal cord produced movements of muscles. It was in the course of exploring the functions of the various parts of the nervous system that he was led to perform a very curious and instructive experiment in nerve suture. He cut the nerves of a cock's wing as they make their exit at the shoulder and sutured the trunk which passes along the flexor aspect of the wing to the root of the extensor nerve; while the nerve which supplies the extensor muscles he sutured to the root of the flexor trunk. At first the wing drooped and was paralytic, but in time motion and control returned. Thus Flourens introduced suture of nerves as a technical procedure; he demonstrated not only that a divided nerve will heal, but it can be made to form an alliance with a strange trunk. And, further, that in time the central nervous system can utilise the novel nerve paths as part of the machinery for executing old and familiar movements. Yet, like Swan's discoveries, those of Flourens—so far as practical surgery was concerned—fell dead from his active brain. Seventy years were to pass before surgeons again discovered the methods of nerve exchange and nerve reimplantation which Flourens had hit upon in his experimental laboratory. Why that should have been so is a puzzling circumstance, for there never was a writer who thought more clearly and expressed himself more emphatically and ostentatiously than did Flourens.

The enigma of the immediate union of divided nerves was not solved in hospital wards, but in a physiological laboratory. At the time when Paget was lecturing in the theatre of the Royal College of Surgeons, Augustus Waller, a young general practitioner waiting for patients in the Kensington district of London, was quietly making observations which were destined to throw an altogether new light on the nature of nerve fibres and nerve union. Waller was an uncommon man with an unusual history. His father, a farmer near Faversham, Kent, had, in a search for health, moved his home to the south of France—to Nice; and hence it happened, when his son Augustus chose Medicine as a profession, he selected Paris as the scene of his studies. He took his medical degree there in 1840, when he was 24 years of age, being thus Paget's junior by two years, and in 1842 began the vigil of a young practitioner amongst the squares of Kensington. At the point we take up Waller's history—in 1849—Richard Owen, the Conservator of the Museum of the Royal College of Surgeons, had just communicated to the Royal Society a research made by the young practitioner in Kensington, on "The Minute Structure of the Papillæ and Nerves of the Frog's Tongue." From that paper we learn that in the final year of his study in Paris, Waller had been caught in the great "cell-doctrine" which at that time swept the dissecting rooms of Europe. He became acquainted then with the splendid opportunities which the transparent tongue of the living frog offers for microscopic studies and discoveries. And now, seven years later, we find him again exploring the frog's tongue to discover the mechanism of taste—the manner in which the terminals of the nerve of taste are brought in contact with the sapid elements of meat and drink. Looking through the microscope he could trace clearly the individual nerve fibres along the tongue to their termination in the papillæ. But the tongue of a living frog, spread out beneath the objective lens of a microscope, although transparent, is also a restless object, and the observer had to adopt means to keep it quiet. Finally, Waller found that the much vaunted and newly introduced anæsthetic—ether, accomplished his object best; but he had tried other means, and although he does not explicitly tell us so, we have no doubt section of the nerves to the tongue was one of them. Hence his purely anatomical paper of 1849 is followed by a short, but much more important paper in 1850.[1] He observed that the terminal fibres going to the papillæ, derived from nerves which he had divided three days before, were undergoing a change of a remarkable nature. The axis cylinders of the fibres and

[1] "Experiments on the section of the Glossopharyngeal and Hypoglossal Nerves of the Frog, and Observations on the Alteration produced thereby on the Structure of their Primitive Fibres," *Phil. Trans.*, 1850, vol. cxl. p. 423.

the myelin sheath which surrounds them, and which are seen distinctly in the transparent tongue of the frog, were breaking up, curdling and becoming mixed together; by the sixth day disorganisation became marked, and by the tenth day a granular pulpy substance replaced the axis cylinder and its myelin sheath. He noted that the change ceased at the point of section; the nerve fibres which lay within the nerve trunk on the proximal side of the section retained their normal outlines; the change stopped short at the point of division, whether it was a motor nerve such as the hypo-glossal that was divided or a sensory nerve like the glossopharyngeal. Now at the time Waller made this observation nerve cells were known and so were nerve fibres, but the relation of the cell to the fibres was not definitely determined, as we shall see, until a much later date. From his experiments, Waller drew the conclusion that the moment nerve fibres were separated from the central nervous system, the part beyond the point of division underwent that curious degenerative change which he was the first to observe.

Waller's observations did not cease with the discovery of the degeneration which goes by his name; at the end of nine months, from the date of section, he discovered that normal nerve fibres were again to be found in the area of the tongue which had been the site of degeneration. He noted the process of regeneration and saw that it commenced, immediately after section, at the cut end of the proximal or central part of the nerve trunk; he saw new and delicate fibres grow out from the cut ends of the old fibres and gradually invade and replace the degenerated fibres. Clearly there was no immediate union in the nerves of the frog's tongue. He went further: he noted that the process of nerve regeneration was assisted by youth and by warmth; repair took place more rapidly in young than in adult frogs, and in the summer time than in the winter time. He tried, too, the effects of daily applications of the galvanic current and also of movement, but did not perceive that such treatment in any way accelerated the outgrowth of the new axis cylinders.

Thus the basal law which must regulate the action of

every orthopædic surgeon in the treatment of nerve injuries of the human body was discovered by a medical practitioner while, in his odd hours, he was searching in the frog's tongue for the manner in which sapid substances are brought in contact with the nerves of taste. Augustus Waller had an eye which was quick to detect the open sesame among the appearances presented to him in the course of his experiments; he was fully convinced he had discovered a key which would open the ravelled nerve tracts of the central, as well as of the peripheral nervous system. Regeneration picked out the course of the most intricate nerve path. In the following year Türck demonstrated that Waller was right; the pyramidal tracts, when they had been torn through by an apoplectic hæmorrhage, underwent a degeneration exactly similar to that which had been observed to occur in the nerves of the frog's tongue. Here we part with Waller; after his greatest discovery he abandoned Kensington and practice and became a scientific Bedouin in the laboratories of the Continent, dying at Geneva in 1870, at the age of 54. The secret of the open sesame by which he gained access to the inner recesses of Nature's palaces became the heirloom of his son—the distinguished Director of the Physiological Laboratory of the University of London.

The events we have been considering centre round the year 1850, and the evidence they yielded came from two sources—divided nerves of the human hand, and divided nerves in the frog's tongue. We are now to pass on to 1880 to examine evidence from a new source—the microscopic examination of sectioned embryos. In that year Wilhelm His, Professor of Anatomy in the University of Leipzig, then in his fiftieth year, published the first part of *The Anatomy of the Human Embryo*, and Francis Maitland Balfour, a young Cambridge "don" of 29, issued the first volume of his *Treatise on Comparative Embryology*, the second volume appearing in the following year, 1881. In ten short years this scion of a house, which hitherto had produced mere statesmen, created, with his own hand and his own brain, a new department of knowledge. Arriving at Trinity College, Cambridge, in 1871 as a lad of 19, he had the good fortune to fall under the spell of Michael Foster, who set him to investigate the embryology of the chick. We have seen that Waller found a source of new knowledge in the frog's tongue; the developing embryo of the dog-fish was to become a mine of discovery for young Balfour. Later experience has proved that the march of events which pass so obscurely as the systems of the body are unfolded in the embryos of higher vertebrates, occur in the developing dog-fish in a vivid and diagrammatic sequence. Balfour, when he commenced his study of the dog-fish embryo, soon recognised that he had hit upon a sure key to the puzzles which confronted those who were trying to unravel the developing processes in man and other high vertebrates. But for that fatal accident on the Alps in 1882, the story of comparative embryology in England would have run a different and a speedier course.

The particular part of Balfour's work which concerns us at this moment is his discovery of the manner in which nerves arise in the developing embryo. He found that a nerve—taking an intercostal as an example—arose from the embryological rudiment of the spinal cord as two colonies or companies of nerve cells, one to form the anterior root and another to form the posterior root, posterior root ganglion and sensory fibres. The cells, as they migrated outwards into the body-wall, laid down the nerve fibres; as the company of cells pushed on towards the ultimate destination of the nerves, some of their members fell out, became elongated, and, joining with neighbours, were thus converted into a nerve fibre. Balfour had no doubt that a nerve fibre was formed by the union or fusion of individual nerve cells. "The cellular structure of embryonic nerves," he wrote, "is a point on which I should have anticipated that a difference of opinion was impossible. He could not understand how His and Kölliker could have come to a totally different conclusion.

It would be difficult to conceive a greater contrast than that which existed between the brilliant amateur, Francis Maitland Balfour, and William His, Professor of Anatomy in the University of Leipzig — the professional student — the founder of Human Embryology. He was a Swiss, born at Basel, and called to fill the chair of Anatomy there in 1857 at the age of 26. He commenced the study of the developing chick and enlisted the services of the medical men of his native town to gather for him specimens of the earlier stages in the developing human embryo, for of these stages medical men had no exact knowledge. When he was called to Leipzig in 1872 he carried with him his lovable ways, his studious habits, and his collection of embryoes, and in 1880 began to publish a systematic account of his researches into the early stages in the development of the human body. He was not the first man to observe that the fibres which constitute the anterior root of a spinal nerve arose as outgrowths or processes of nerve cells situated in the ventral region of the embryonic spinal cord, but he was the first who had worked through early stages in the development of the central nervous system from end to end, and found in every part of it that nerve fibres arose in the same way—namely, as outgrowths from nerve cells. He therefore laid down as a law that every nerve fibre was the outgrowth—was a process or part of a single nerve cell. For this reason it was as natural for a nerve fibre to degenerate when it was separated from the cell of which it formed a part, as it was for a branch to wither and die when cut off from a tree. He admitted that the outgrowing fibres were surrounded by a company of cells derived from the central nervous system; but these, he held, were merely accessories or nurses designed for the protection, perhaps for the nourishment, of the growing fibres; they became the nuclei of the sheath. Wallerian degeneration, it must be admitted, found a better explanation from the observations of His than from those of Balfour.

At the beginning of the present century both His and Balfour had their followers.[1] If His was right, then the distal part of a cut nerve was merely a vital scaffolding into which the budding fibres grew from the proximal part; but if Balfour was right, the part of the nerve beyond

[1] See *The Healing of Nerves*, by Charles A. Ballance and Purves Stewart, London, 1901.

the point of division was much more than a mere scaffolding; it was a potential nerve tract. The sheath cells, being the representatives and descendants of the neuroblasts, which laid down the original nerve fibres, should be able to resume their original function, again becoming neuroblasts and formers of new nerve fibres. In 1904, Dr. Ross G. Harrison, a young graduate of the great school of anatomy which Franklin Mall had built up in Johns Hopkins University, Baltimore, stepped in with a series of experiments which seemed to prove that the outgrowth theory of His was the truth. On this occasion it was the developing tadpole that yielded the crucial evidence. The emigrant colonies of cells which pave the path of the outgrowing anterior or ventral nerve roots, have their original home along the dorsal crest of the embryonic spinal cord. Harrison removed the posterior part of the dorsal crest from developing tadpoles, and saw that, in the region deprived of the crest, the fibres of the anterior roots grew out as usual, but they were sheathless and unaccompanied by the usual emigrant cells. These and other experiments seem to settle the question of the origin of nerve fibres as outgrowths from single cells. A nerve fibre does not arise by a combination of cells, but is the outgrowth or process of a single cell.

We now return to the puzzle of the apparent immediate union of a divided median nerve as propounded by Paget. To understand how the matter developed we have to know something about Jean Joseph Emile Létiévant. In 1867, when holding a combined anatomical and surgical post in l'Hôtel-Dieu of Lyons—in which city he was born thirty-seven years before—he cut the median nerve high in the arm of Joseph Gaillard to arrest an attack of tetanus, the result of a gash in the palm. We suspect that Létiévant was rather glad to have the opportunity of performing that operation for the following reason. Three years before, M. Lengier, Professor of Clinical Surgery in Paris, had created a sensation by announcing that he had sutured a median nerve which had been divided above the wrist, and that feeling and motion had returned in the area supplied by the nerve almost immediately. Nélaton reported that he had observed a similar result in a case performed in the preceding year. Paget's doctrine of immediate union was thus regarded as established. Létiévant's knowledge of anatomy and physiology made him certain that an error of observation was being made by his colleagues in Paris, and it would be expecting far too much of his eminently human personality to suppose that he would allow to pass by unused an opportunity of exposing the fallibility of his metropolitan rivals. He observed in Joseph Gaillard seven hours after the operation, when the ends of the median nerve were still apart, that there was both feeling and movement in the area supplied by the median. As the cut ends of the nerve were still apart, it was plain that the sensation in the hand was not due to immediate union. But he also noted the extent and kind of feeling that was present. There was one small area on the palmar aspect of the index finger where there was no sense of pain even when a pin was pushed deeply into the flesh, but outside this anæsthetic area there was a wide field extending almost to the limits of the anatomical distribution of the median where pressure, or rubbing by means of a paper arrow, was distinctly felt, where a pin-prick gave rise to pain, where very cold or very hot things were recognised, but where the touch of a feather, or slighter degrees of heat and cold were not felt. In this area of blunted sensation the two points of the compass, however widely separated, were felt as one. Nine months later Joseph Gaillard, who had returned to work, reported that he had recovered the use of his hand. On visiting and examining him Létiévant found that both sensation and movement were in reality exactly as they had been after the operation nine months before. Fourteen months after the operation normal sensation and movement began to return to the median area, and in nineteen months recovery appeared to be complete. It was quite evident, so Létiévant inferred, that the feeling which Paget, Langier, and Nélaton had observed in the hands of their patients after section of the median nerve was due, not to immediate union of the cut ends, but to another circumstance altogether, namely, that the area of the median was not only supplied by that nerve, but had also an accessory or supplementary supply. Every nerve area, so he postulated, must have a supplemental power which depended on two circumstances: (1) the anastomosis between neighbouring nerves, as between the median and ulnar, and median and radial; (2) the power which the papillæ of the skin have of feeling stimuli applied, not directly to them, but at a distance. "For instance," said he, "I take the finger of Mr. X. between my index and my middle finger, and turning my head as I fix my attention on my hand, I ask Mr. X. to rub *his* finger. I feel the rubbing; I can recognise the site, intensity, and direction of the rub." That was proof positive that the papillæ of his finger could recognise stimuli applied not directly to them, but at a distance. It is strange that Létiévant should have so completely overlooked the fact, vigorously preached at that time by Duchenne, that all tendons and joints, including those of the hand and fingers, were richly supplied with sensory nerves.

With the publication of his *Traité des Sections Nerveuses* in 1873, Létiévant believed he had killed the theory of immediate union of nerves, but thirty years later, in spite of his observations and criticisms and of the discoveries of Waller and of His, there were many surgeons who still believed in it. In the opening years of the present century the theory of immediate union was finally killed and buried at the London Hospital—the hospital which produced Hughlings Jackson, Hutchinson, and Little. Dr. Henry Head was chief executioner. Many years of exact and systematic observation had led him to see that peripheral nerves contained sensory fibres of several different kinds, subserving different functions. He and James Sherren examined systematically every case of nerve lesion which came within the walls of the hospital—particularly cases of the kind noted by Paget, Langier, and Nélaton, and on which Létiévant had founded his theory of "supplementary supply." They saw that before any decision could be reached, it was necessary to discover the extent to which the subcutaneous and deeper tissues had the power of recognising stimuli applied to the surface of the hand or body. One spring day in 1903, Dr. Henry Head appeared in the wards with his arm in a sling; by having his radial (n. cutaneus antibrachii dorsalic) and external cutaneous nerves divided at the elbow and short segments excised at the points of division, he had produced, and carefully mapped out, an area of the skin on the outer or radial side of his forearm and hand. When this nerve-sectioned area of skin was touched with the point of a pencil —the test employed by Paget—Dr. Head not only felt it, but could also tell the point at which the stimulus had been applied. The deep parts were, therefore, supplied with nerves —nerves of deep sensibility—which could recognise and locate the position of pressure stimuli. But the skin itself was insensitive to pin-prick, could not respond to the compass test or detect the light touch of cotton wool or recognise heat from cold. In six weeks from the time of section the power to recognise pin-prick began to return to the skin of the forearm and in 200 days it had extended over the anæsthetic area in the arm; the power of feeling pin-pricks was accompanied by the ability to recognise temperatures below freezing point or above 50° C. Then at the end of a year the affected area of skin began to be sensitive to a new set of stimuli—to the "light touch" of cotton wool and to lesser degrees of heat and cold, and by the end of two years the whole area had regained its normal power — discrimination of the compass points being the last faculty to return. From that experiment and from many other observations, Dr. Henry Head drew the conclusion that peripheral nerves carry at least three distinct sets of fibres: (1) those which supply sensibility to the deep parts—to muscles, tendons, joints and bones; (2) a set which endowed the skin with a low or "protopathic" sensibility —the power to recognise painful stimuli and stimuli produced by the application of extreme degrees of heat and cold; (3) a set which supplied the skin with an "epicritic" sensibility — light touch, slighter degrees of heat and cold, and the discrimination of compass points.[1] In cases where

[1] See Dr. Henry Head's "Marshall Hall Address," *Brain*, 1905, Part CX. p. 9; Henry Head and James Sherren, "The Consequences of Injury to the Peripheral Nerves in Man," *Brain*, 1905, Part CX. p. 116; James Sherren, *Injuries of the Nerves and their Treatment*, London, 1908.

the median nerve was cut, it was found that epicritic sensibility was lost over the area of anatomical distribution of the nerve. Therefore the epicritic supply for this area was contained entirely in the median nerve. But it was otherwise as regards the protopathic supply ; it was lost over a comparatively small extent; therefore it was clear that the median area derived a protopathic supply from neighbouring nerves as well as from the median. When the median was cut at the wrist the tendons and joints of the fingers still retained their nerves of deep sensibility from muscular branches supplied by the ulnar nerve in the hand and the median in the forearm, and the fingers therefore remained sensitive to deep pressure or to rubbing—vibration—the nature of the stimuli applied by Létiévant. With the publication of Head's discoveries there was no longer any need to believe in the theory of immediate union of nerves.

Every step taken by a surgeon to restore continuity to a nerve must be based on (1) the discovery by His that a nerve is an outgrowth from, and part of, a nerve cell; (2) the discovery made by Waller that the part of a nerve fibre lying distal to the point of division always dies and can only be replaced by an outgrowth from the end of the fibre which lies proximal to the point of division; (3) on the discovery made by Flourens, elaborated by Philipeaux and Vulpeau in Paris in 1863 and onwards, and later, in Cambridge, by Langley and Anderson, that if the proximal end of one nerve is sutured to the distal end of another, the fibres of the proximal nerve will invade and gain dominion over the area of the distal nerve; (4) the path of outgrowing nerve fibres must be paved with living sheath cells.

These discoveries have been utilised in many different ways to restore movement to groups of paralysed muscles. In 1895, Ballance sutured the trunk of a paralysed facial nerve to the proximal end of that part of the spinal accessory which supplies the muscles of the neck and shoulder. In time the messages which formerly passed to the shoulder began to reach the muscles of the face. The principle represented by that operation has been extended to many parts of the body.

To mend a wide gap in a torn or divided nerve, a bridge or scaffolding of a very definite kind has to be inserted before the budding nerve fibres from the proximal side of the lesion can be induced to cross and so reach the distal trunk. The scaffolding must supply sheath cells of a living nerve—a nerve derived from the patient's own body or from the body of another man or woman. In 1876, Albert attempted to mend a gap in the median nerve by inserting a segment cut from the tibial nerve of an amputated limb—a procedure which Mayo Robson successfully carried out in 1888. From a theoretical point of view the segment of nerve which is introduced to make good a gap, should be composed, not of intact, but of fibres which have already undergone degeneration in consequence of division. It is a degenerated, not a normal nerve which affords a suitable scaffolding or bridge for the fibres which grow out from the distal end of the injured nerve trunk to repair a gap or breach. Hence it is a nerve which has already undergone degeneration that we should use as a nerve-graft.

We have seen that Joseph Swan discovered the well-known fact that part of a motor nerve may be cut through and yet there may be no *apparent* loss of movement in the parts supplied by that nerve. This fact has also been utilised and is capable of extension. A partial section of a healthy nerve trunk sets free a number of proximal nerve fibres to which a paralysed nerve may be attached and into which the budding fibres from the proximal or parent trunk will grow and in time conduct messages from the central nervous system. Thus we see that our orthopædic practice, so far as the treatment of nerve injuries is concerned, rests on a century of laboratory experiment and clinical observation.

CHAPTER IX

THE INTRODUCTION OF TENDON TRANSPLANTATION

In this chapter transplantation of tendons is discussed as a measure which the surgeon may adopt if nerve suture fails. The idea of transplanting the tendon of a healthy muscle so as to perform the work of one which is permanently disabled, dates back only to 1881, while Vanghetti's idea of utilising the muscles of stumps as the source of motor power for artificial limbs is even more recent, having been first put into practice in 1896.

THE human body is a machine built up of a complex series of levers which are driven by engines of a peculiar type. We have seen that these engines are under the direct control of certain nerve cells—" driver " cells—situated in the spinal cord, and that the drivers are connected with the engines by a prolongation of their bodies, which, for the sake of continuing the simile, we may call an arm, and its end plate, a hand. When these arms are cut or broken by a gunshot wound the engines to which they reach are brought to a standstill. In the previous chapter we have been examining the conditions under which the lopped-off arms of the driver cells may be reproduced. We found that if a continuous path of sheath cells could be provided and the engines kept in a true state of rest, we have every reason to expect that the missing arms will again reach and control their muscular machines. At the same time the arms of the signaller cells, situated in the posterior root ganglia, must also reach their stations in the paralysed muscle, for without a due and prompt supply of signals the driver cells can exercise no effective degree of control.

If, however, our efforts to establish a nerve path to a

muscle fail, or if nerve repair is an impossibility, what are we to do? Let us take the case where the musculo-spiral nerve has been destroyed by an extensive wound in the region of the lower part of the humerus. All the muscular engines which extend the wrist, the thumb and fingers, become useless and Nature proceeds to scrap them. The wrist becomes permanently "dropped" and the fingers fixedly clinched, unless we restrain their healthy opponents situated in the flexor side of the forearm. Here, then, we have a forearm furnished with ruined engines on one side and provided on the other with a superfluity of machines in prime order but rendered useless because they have no opponents. Unless we can provide the healthy flexor engines with antagonists Nature will ultimately scrap them too. One would not think that it required a high order of genius to perceive the possibility of yoking the traction piston of the destroyed extensor engines to some of the superfluous healthy flexor engines, thus re-establishing a balanced action at the wrist and in the fingers, but, as we shall see, such a transfer is one of the latest discoveries in orthopædic surgery.

In the beginning of 1881 there came to the clinic of Professor V. Dumreicher, in Vienna, a lad, Joseph N., aged 16, who had suffered in infancy from infantile paralysis, which ultimately left the muscles of the calf of the right leg, and the deep flexor muscles under them, hopelessly paralysed, while the muscles in front and on the extensor aspect retained a full degree of action. The result was that the foot had become dorsi-flexed, and walking was done by applying the limp heel to the ground. The case fell for study and treatment, as others of the same kind did, to Dr. Karl Nicoladoni, assistant in the clinic, a native of Vienna, and at that time on the point of being transferred to Innsbruck as Professor of Surgery. In 1881 he was already a man of 34 with a long experience of orthopædic cases. He was interested in the surgery of tendons,[1] being an advocate of their careful suture in all wounds—particularly those of the wrist and hand. In the case of Joseph N., Dr. Nicoladoni

[1] "Ein Vorschlag zur Sehnennaht," *Wien. med. Woch.*, 1880, p. 1406.

conceived the idea of redressing the deformed and unbalanced foot by yoking the healthy peroneal muscles to the heel.[1] Hence on the 15th of April 1881 he raised a triangular flap of skin from the outer side of Joseph's leg and ankle, thus exposing the tendons of the two peroneal muscles in the malleolar region and the tendo Achillis for some distance above the heel. He divided the peroneal tendons on the outer side of the foot, separated them from the fibula for some distance and then drew them across towards the heel and sutured them within a cleft in the tendo Achillis. The wound did not heal up satisfactorily; and, unfortunately, while the case was still under treatment, he had to take up his new post in Innsbruck. At first the peroneal muscles appear to have answered Dr. Nicoladoni's anticipations, but latterly they broke down, and in the light of recent experience we can guess the cause of the failure. Muscles at any time are really very delicate machines, and those which have undergone the severe damage that necessarily attends transplantation need protection from force and stress of all kinds and the most careful nursing for a long period before they are fit for daily use. The cause of the failure in Dr. Nicoladoni's first case may be attributed to the want of after-treatment.

As so often happens with the efforts of the pioneer this well-conceived experiment by Dr. Nicoladoni passed unnoticed. So little attention had it attracted in the surgical world that when in 1892 Dr. B. F. Parrish of New York performed an operation of a somewhat similar kind, he had every reason to believe that he had introduced a procedure which was new to orthopædic surgery.[2] Dr. Parrish's operation was carried out, 15th May 1892, on a girl almost 4 years of age—eleven years after Nicoladoni's. In her first year the patient had been the subject of infantile paralysis, which left both her anterior and posterior tibial muscles useless and the arch of her foot flattened. Dr. Parrish observed that although

[1] "Nachtrag zur Pes Calcaneus und zur Transplantation der Peroneal Zehnem," *Arch. f. klin. Chir.*, 1882, Bd. xxvii. p. 660.

[2] "A New Operation for Paralytic Talipes," *New York Med. Journ.*, 1892, vol. lvi. p. 402.

the tibialis anticus was paralysed the muscle which lies on its outer side—the extensor of the great toe—possessed a full degree of activity, and he therefore conceived the idea of yoking the active to the passive tendon, so that the power which should have been spent by the extensor longus hallucis on the great toe would be exerted on the inner side of the flattened tarsal arch. He therefore made a vertical incision in front of the leg and above the ankle, thus exposing the active and passive tendons where they lie almost side by side. He freshened the adjacent borders of the tendons and sutured them together and enclosed the leg and foot in a bandage of plaster of Paris. The result was successful, and at the conclusion of his account of the case Dr. Parrish shows us that he had appreciated to the full the wide application of the principle on which he had based his operation. "However," he wrote, "the most important principle of grafting tendons and having a live muscle to do the work of a dead one is that which I wish to particularly establish in this article."

In the last decade of the nineteenth century orthopædic surgeons took much to heart the sad case of the children who had suffered from infantile paralysis and sought by every means in their power to remedy their deformities. Precautions to prevent weakened muscles from being overcome by overstrong healthy opponents and perseverance in re-education of muscles proved to be really efficient means, but nerve suture, and particularly tendon transplantation, as devised by Nicoladoni and by Parrish, were found to be of great assistance in many cases. In 1897 Bradford [1] of Boston gave an account of twenty-seven operations in which he had transplanted tendons —"tenoplastic" operations, as he named them; in 1899 Professor Vulpius [2] of Heidelberg was able to find published records of thirty cases; in 1899 Sir Robert Jones and Mr. Tubby gave an account of fourteen operations in which they had successfully transplanted tendons; in 1908 Bradford

[1] Ed. H. Bradford, "Tenoplastic Surgery," *Annals of Surgery*, 1897, vol. xxvi. p. 153.

[2] *Trans. Amer. Ass. Orthop. Surg.*, 1898, vol. xi. p. 439.

and Souttar gave a list of 500 cases. Thus we see that once the operation of tendon transplantation was introduced it spread like wildfire through the orthopædic establishments of the world.[1]

It promises to be equally serviceable in military orthopædic hospitals. As experience of the operation increased, modifications were introduced. Instead of exposing an extensive field of operation, as was originally done by Nicoladoni, well-placed incisions, of a button-hole type, made over the sites of tendon-section and suture were made to serve the needs of the operator. A new course was made for the transplanted tendon by tunnelling in the tissues from one incision to another. Devices were employed to prolong the transplanted tendon—if it proved too short. Professor Lange of Munich introduced leashes of silk ligature for this purpose. It was found that the transplanted insertion did better when grafted to the bone and periosteum than when sutured to the tendon stump of a paralysed muscle. We see an example of direct implantation in the operation which Mr. Tubby devised and carried out in 1901 to restore the power of supination in a case of spastic paralysis of certain muscles of the arm; he separated the insertion of the pronator teres from the flexor aspect of the radius, brought the muscle through the interosseus space and implanted it on the dorsal or extensor aspect of the radius, thereby converting it from a pronator to a supinator of the forearm.

Experience in the dissecting room leaves the student with an impression that muscles are hardy structures and capable of being easily shifted. The truth is exactly the opposite; muscles are amongst the most delicate, most highly organised of all the tissues of the body, and to move them successfully great judgment and patience are required. Every bone in the body is shaped so as to serve the functions of its situation; if it become displaced in its relationship to other bones or changed in shape,

[1] The early history of tenoplastic surgery may be found in Mr. A. H. Tubby's *Text Book of Orthopædic Surgery*, 1912, and more fully in *The Treatment of Infantile Paralysis*, by Oskar Vulpius. Translated by A. H. Todd, 1912.

even to a slight degree, its living cells, the osteoblasts, at once commence to remodel and adapt it to serve its functions under the altered conditions. That truth, as Mr. Morley Roberts has pointed out, is equally applicable to muscles. Every fibre of a muscle is set, fixed, and arranged according to a normal line of action ; if we change the line of action of a muscle then every fibre of that muscle has to undergo a readjustment to answer the new conditions of work ; the more we alter the line of action the greater burden do we throw on this power of adaptation which is inherent in every muscle. If the line of action is altered too much, a point is reached when adaptation becomes impossible and degeneration sets in and the muscle is reduced until it becomes a mere ligament. The degree to which the line of traction is altered must be made a matter for consideration in every operation of tenoplastic surgery. Let us take an example—a case where we wish to make the peroneus brevis serve in place of a paralysed tibialis anticus. The fibres of the peroneus brevis are arranged to act in a line of traction represented by the axis of the fibula. If the tendon of the peroneus is brought across the front of the ankle and planted in the inner side of the arch of the foot, then the line of traction is so changed that its fibres have to undergo a rearrangement to suit the new conditions—one which is probably beyond their adaptative power.

Another principle we have to keep in mind is that of balance. Muscles are reciprocal engines which are only effective when rightly coupled. If we set a 100 h.p. engine against a 1 h.p., unless we give the latter a due advantage in leverage, the weaker engine will be quickly destroyed by the stronger. Let us take the case of a soldier whose external popliteal nerve has been destroyed beyond repair and in whom the dorsi-flexors of the foot are paralysed. With a view of replacing them we bring the tendon of the tibialis anticus through between the tibia and fibula, suture it to the dorsum of the foot and thus convert it into a dorsi-flexor of the foot. Even if our operation is successful we are not likely to attain our ultimate aim, because the weak muscle thus transplanted cannot withstand the overwhelming strength of the muscles of the calf against which we have pitted it. In all tenoplastic operations we must try to secure a transplant which is sufficiently strong to serve as an antagonist.

In designing a tenoplastic operation the nourishment—the blood-supply of the muscle and tendon—becomes a matter of first importance. How far a transplanted tendon has ultimately to undergo a reorganisation of its substance in consequence of damage done to its circulation by the operation we have no certain evidence, but from what we know of bone and other grafted tissues it is possible that to a certain extent it may only serve as a scaffolding on which a new tendon is organised. The blood-supply of the part of the tendon adjacent to the muscle will likely remain undamaged so long as the vessels of the muscle itself remain intact. Of much more importance than the blood-supply of the tendon is the supply of the muscle itself; if once the myogenic cells of a muscle or part of a muscle are killed by a "blood starvation," death of that muscle or that part is final and complete. To trace the beginning of our knowledge concerning this "ischæmic" disaster which sometimes overtakes muscles, we have to visit Halle during the years which followed the close of the Franco-Prussian War. Richard von Volkmann, who had served as consulting surgeon to several armies during that war, had then returned to his professorial chair, a man of 41. As an orthopaedic surgeon, he was particularly interested in the surgical disorders of muscles. He noted that, in cases of fracture of the forearm or of excision of the elbow, a distinctive form of contracture of the muscles of the forearm was apt to occur. The contracture was such that in a few weeks the wrist became acutely flexed and fixed and the fingers clenched "until even the nails were drawn into the flesh of the palm." He found the contractured muscles became as resistant as if they had been composed of sodden leather; he observed that neither the administration of chloroform nor the application of force could undo the deformity of the hand. He also noted that the contracture usually appeared in fractured forearms which had been bound too tightly in splints, and suspected that an interruption of the blood-supply to the affected muscles was the cause of their contracture. The breaking up of the contractile substance in these muscles resulted, he believed, from the deprivation of oxygen consequent on the arrested blood-supply from compression. He inferred that contractured muscles were muscles which had been asphyxiated, killed, and therefore had undergone a "rigor mortis" within the living forearm. The connective-tissue elements, which survived, then underwent a reaction and multiplication, for they had to deal with the masses of dead myogenic substance which surrounded them. When a muscle of a living animal was shut off from its circulation by ligature of its vessels for a period of six hours, Von Volkmann found that disintegration of the muscular substance took place—varying in its intensity according to the length of time and the extent to which it had been deprived of its blood-supply. Further experience has proved the truth of Von Volkmann's conclusions. In every tenoplastic operation the blood-supply of the muscle must receive the most careful protection. A muscle is no more capable of contracting if deprived of its blood-supply than an internal combustion engine is capable of working when its stream of aerated petrol is cut off. Nor can we conceive the possibility of transplanting or grafting a muscle or part of a muscle because we have no means at our disposal of ensuring an efficient supply of blood to a muscle graft. If such a graft or transplant were to hold, only its shell, not its contractile substance, would survive.

It is not necessary to mention the safeguarding of the nerve supply of the re-inserted muscle—its triple nerve supply of motor, sensory, and vascular nerve fibres. Neither muscle fibres nor nerve fibres must be subjected to stress during the healing period that follows a tenoplastic operation. If there exists a deformity, that deformity must be reduced before the tendon is transplanted. The deformity must be over-corrected so that the transplanted tendon and its muscle may be maintained in a relaxed condition while the damage from the operation is repaired.

Professor Lovett of Boston has rightly emphasised the fact

that the tenoplastic operation itself is a minor part of a surgical campaign which aims at restoring movement to a joint. The process of re-education is the most important phase of the treatment. Suppose, for example, that we are dealing with a case of hopeless paralysis of the muscles supplied by the musculo-spiral nerve in the forearm, and that, to provide the power of extension to the fingers, we have yoked the flexor carpi ulnaris to the extensor tendons of the digits. Hitherto the central nervous system of the patient has been accustomed to deal with the flexor carpi ulnaris—to receive messages from it and dispatch orders to it—as part of the mechanism concerned in wrist movements and in wrist fixation. Formerly, when the fingers were extended the flexor carpi ulnaris came into action as a fixator of the wrist, but after transplantation it has to play the part of active extensor of the fingers. Before the operation, the brain had to drive it as a member of the flexor team of muscles; now it has to drive it as a member of the extensor team, which requires a readjustment—a re-education of the central muscle-driving mechanism. Re-education demands not only patience, perseverance, but also intelligence on the part of the patient—requires months of repeated trial and failure before proficiency begins to come. The successful transplantation of a tendon depends on the skill of the surgeon, but the successful restoration of movement, which is the aim of every tenoplastic operation, depends on the intelligence and will-power of the patient.

Nature has shown remarkable ingenuity in providing means to obviate friction during the action of a muscle. The muscle itself is slackly fixed within its sheath, its fibres are set so that they can glide on each other; the tendon is surrounded by loose connective tissue or by a synovial sheath. The injury which attends the transplantation of a tendon necessarily leads to the formation of adhesions and some degree of fixation. Hence the importance of commencing the process of re-education and of movement of the tendon as soon as possible after the operation.

Closely allied to tenoplastic operations are those performed on the stumps of amputated limbs in order that muscles which would otherwise waste and disappear may be utilised as a source of motor power for artificial limbs. Transplantation of tendons is, as we have seen, a procedure initiated by Nicoladoni in 1881, but the "kinematisation of stumps" is even of more recent origin, dating back only to 1896. The second Italian campaign in Abyssinia was then at an end, the disastrous battle of Adowa had been fought, and many Italian soldiers who had lost a limb in the savage rigours of that campaign were returning home to be fitted with artificial substitutes. In many cases the Abyssinians mutilated their prisoners by cutting off the right hand. Now, any substitute for a normal limb, unless it be a bucket and stump fitted as a walking support, must be capable of performing certain movements, and these movements must be under the control of the wearer. Dr. Giuliano Vanghetti, an Italian physician, wishing to do something to ameliorate the condition of his hapless countrymen, conceived the idea of using the muscles of the stump as a source of driving power for artificial limbs—particularly as the moving force for an artificial hand. In the forearms of these men there were still the intact muscles, formerly the means of winning a livelihood, but now useless because they had been rendered leverless. The problem which this Italian physician had to solve was not one of inventing appropriate and simple forms of digital levers, but the much more difficult one of devising a means by which the muscles could be yoked to such levers. His first idea was to isolate the distal and tendinous part of a muscle and enclose them with a covering of skin—thus producing a retractile muscular polypus situated on or near the terminal end of the stump. It was to the neck of such a peg-shaped, muscular mass—a clava, as he named it—that he proposed to attach the cord by which the artificial finger was to be moved. Two or more clavæ could be elaborated on the same stump. Dr. Vanghetti's next step was to unite the free ends of two of these clavæ, thus forming a muscular loop or ansa, through which the driving cord could be drawn.

Dr. Vanghetti was a physician, and therefore had no opportunity of applying his idea, nor could he prevail upon any surgeon to put his plan into practice. Hence in 1898[1] he published a work setting out the theory of such operations, but it failed to attract the attention of orthopædic surgeons. In 1899 he again published his idea, which had developed in the meantime as the result of experimental investigations. On this occasion he was fortunate enough to enlist the interest of Professor Ceci, who, in 1905, was able to publish an account of the first case in which the Vanghetti operation had been applied. Professor Putti did his first kineplastic operation in 1911; by the end of 1914 about twenty patients had been operated on, mostly in Italy. Then with the outbreak of war the technique of the operation was further elaborated by Professor Sauerbruch of Zurich and by Italian surgeons. It was Sauerbruch who introduced the method of tunnelling the movable cap or "plastic motor" of the stump.

One basal principle Dr. Vanghetti had to observe: a muscle, or a group of muscles, can act only if provided with an antagonistic force. If a muscle contracts it can only act again if it has been elongated—either by gravity, or better, by the counteraction of an opposing muscle or group of muscles. He had to design a method of coupling the extensors and flexors of the forearm so that they could act as antagonists to each other; that was ultimately done by uniting them in the stump and making the cap of the stump movable. If short distal segments of the radius and ulna were separated and included in the cap of the stump, so that a false form of wrist-joint was created between the cap and the forearm, then the muscles of the forearm were provided with a means of producing opposed flexor and extensor movements. The problem of obtaining a plastic motor or movable stump-cap having been solved, the next difficulty which had to be overcome was the fixation of the driving cords—the artificial tendon of the digital levers—to the cap. That, as we have

[1] For an account of the early history of kinematisation of stumps, see Dr. Vanghetti's "Progressi Attuali Della Plastica Cinematica," *La chirirgia degli organi de Movements*, edited by Dr. V. Putti, 1917, vol. i. p. 71. See also lecture by Dr. Putti, *Brit. Med. Journ.*, 1918, vol. i. p. 635. Also an account of the operation given by a surgeon in the *Brit. Med. Journ.*, 1918, vol. ii. p. 68.

seen, can be accomplished in one of three ways—by forming a peg or clava; a loop or ansa; or by tunnelling the muscular mass and then lining the tunnel with skin. By means of a bit inserted in the dermal tunnel, the driving cords of the digits can be yoked and driven. As in tenoplastic operations, the patient has to pass through a long and tedious process of re-education.

The idea of kineplastic surgery which Dr. Vanghetti first conceived in 1896 did not find favour with orthopædic surgeons. But with the outbreak of war, conditions suitable for the application of a Vanghetti operation became only too common. Dr. V. Putti, Professor of Orthopædic Surgery in the University of Bologna and director of the Rizzoli Institute of that city, developed and improved the original methods devised by Dr. Vanghetti. In the commencement of 1918 Dr. Putti had operated on fifty cases, his best results being obtained in arms which had been amputated above the wrist or above the elbow, but he also attained success in the kinematisation of stumps of the thigh.

SIR WILLIAM ARBUTHNOT LANE

CHAPTER X

THE INTRODUCTION OF CERTAIN ORTHOPÆDIC METHODS TO BRITISH SURGERY

In this chapter I have sought to illustrate some of the principles which guide the orthopædic practice of modern British surgeons by selections culled from the writings of Sir William Macewen (born 1848), Professor of Surgery in the University of Glasgow, Sir William Arbuthnot Lane (born 1856), senior surgeon to Guy's Hospital, and Sir Robert Jones (born 1858), lecturer on Orthopædic Surgery, University of Liverpool.

IN this chapter I propose to return to the busy "Borough" Hospital to mark certain phases in the evolution of a surgeon who has exercised, as time will amply prove, a very real influence on the outlook and practice of British surgeons. The period which concerns us is one of fifteen years, stretching from 1873, when William Arbuthnot Lane, at the age of 17, entered Guy's as a student, until 1888, when, at the age of 32, he became assistant surgeon to his hospital—thus bringing the dissecting-room phase of his life to an end. Anyone searching into the dates of his studentship will find that after spending four years in taking the membership of the College of Surgeons he apparently formed a new resolution and spent the next five in satisfying the requirements which the University of London demands of her graduates. There were thus nine years of hard study behind him when he re-entered the dissecting room as demonstrator of Anatomy. We know with precision the exact nature of the problems which engaged his attention as demonstrator, for he has recorded his studies in contemporary numbers of the *Journal of Anatomy and Physiology*, and in the *Reports of Guy's Hospital.* If he had followed the fashion of the time he would have taken up some research of a microscopic or embryological nature, or devoted himself to some problem in comparative anatomy or morphology, but he chose none of these. He harked back to a subject which had come to be regarded as out of date—the manner in which the skeleton of the human body becomes adapted to the particular burden it has to bear in life. His surroundings gave him opportunities of studying the skeletons of brewers' draymen, of shoemakers, of coal-heavers, and of deal-porters, and of noting the manner in which vocation could mould each bone for definite purposes. In his eyes the human skeleton became "a crystallisation of the lines of force." We note, as he publishes his various studies, that no learned list of reference is ever given; there are no allusions to the writings of men who had laboured before him; his own field of observation and inference is sufficient for him. We note, further, a certain degree of impatience in his early papers; he never does show us the ladder of facts by which he has reached certain conclusions; it is enough for him that he has reached them.

Amongst those conclusions there is one which proved of great value to his life's work. We can best illustrate it from his study of the skeleton of a brewer's drayman. The dray-man's spine, he observed, had become moulded to meet its vocational burden—a barrel of beer, poised on the right shoulder—but this vocational adaptation had been purchased at a price. Weight was no longer transmitted equally, but unequally, through its intervertebral joints. On that side of the joint on which the increased pressure fell there was a reaction; the articular edges of the joint had undergone a change. In the opinion of all other observers the articular change noted by Arbuthnot Lane was a manifestation of disease—of chronic rheumatism, but for him it was a functional response—a growth reaction—called forth by the altered lines of force. He noted that the nature and extent of the response or alteration depended on the age of the individual. Under the age of forty a skeleton possesses a positive power of reaction. By throwing out buttresses and undergoing changes in shape, the various bones of younger people can adapt themselves to altered burdens. About the age of 40, the adaptative powers of the skeleton undergo a diminution; the growth reactions of bones become passive in nature; the articular surfaces respond by undergoing not a positive sound growth but a plastic deformation. Above all, he became convinced that the bones of a limb were so exactly shaped and the articular surfaces so precisely moulded for the daily burden they had to carry, that the introduction of the slightest anatomical deformity must upset the mechanism of every part of that limb, and induce adaptative changes in its joints.

In 1888, at the age of 32, Arbuthnot Lane became assistant surgeon to Guy's Hospital. We propose to trace the influence of his dissecting-room experience on his surgical practice, and for that reason pass straight to the consideration of an episode which occurred at the end of 1893. On 17th December of that year a man, aged 34, who had fallen and sustained a fracture of the leg two inches above the ankle-joint, came under his care in Guy's Hospital. Attempts by all the conventional methods of manipulation and modes of traction failed to reappose the fractured ends. The failure cannot have surprised him, for he had by this time come to the conclusion that the setting of fractures, particularly those of the lower extremity, by manipulation, was a "myth"—one of those "fossilised reproductions of text-books." However, in this particular case he determined to give the patient the advantage of the recognised method of treatment and applied a splint; after waiting three weeks, and seeing no sign of repair, he resolved to apply an altogether new method of treatment. And yet not altogether new, for sixteen years before, in 1877, Lister, in commencing the London phase of his great career, had laid open a broken patella and wired its fragments together. "A novel and by some then thought to be an unjustifiable procedure."[1] Nor would the means which Arbuthnot Lane proposed to use have been possible unless he had mastered the secrets which Lister was the first to reveal. The method he proposed to adopt was one to secure exact reapposition of the broken ends of the tibia, because he knew that the ankle-joint was moulded to serve only one shape of tibia—the exact shape which existed in the patient's limb before the accident. If the fragments were not set exactly then the tibia was left deformed and Nature had to work out a new form of ankle-joint. In such a case she would likely have to remodel the joints of the knee and hip as well to suit the new lines of stress. The surgeon, he knew, often left Nature an impossible task; the worse the surgeon, the greater was the burden left to her. His experience of hospital practice had shown him that such accidents, even when treated by surgeons of skill and experience, did often reduce a workman to a state of life-long beggary. He therefore resolved "to treat the bones as one would the broken leg of a table or chair." The operation—osteo-synthesis, Albin Lambotte has named it—was carried out on 8th January 1894; the broken ends of the tibia were exposed, but required both ingenuity, patience, and force to bring them into perfect apposition. Two screws were inserted across the oblique line of the fracture to keep the fragments in place. A week later, on 15th January, two similar cases were admitted to his wards and were treated in the same manner.

Three months later, at a meeting of the Clinical Society of London held on 13th April 1894, Arbuthnot Lane announced his new method of treatment in a brief paper—the text only occupies five pages of the *Clinical Journal*—

[1] *Lord Lister*, by Sir Rickman J. Godlee, Bart., 1917, p. 420.

which he entitled "A Method of Treating Simple Oblique Fractures of the Tibia and Fibula, more Efficient than those in Common Use." He stated that the results obtained by treating fractures of the leg by the customary method of reapposing the broken ends by manipulation and of retaining them in apposition by splints "were disastrous in the extreme" so far as the functional utility of the broken limbs was concerned. "Why should we hesitate to apply common sense?" he asked; "we do so when the patella is broken." At that meeting Arbuthnot Lane not only described a new method of treating fractures, but he enunciated an altogether new surgical standard. He saddled the surgeon, not Nature, with the entire responsibility of obtaining a perfect result in the mending of broken bones. He knew very well from his dissecting-room studies that Nature had an almost unlimited power of adapting living structures to new conditions; but why should the surgeon by his carelessness throw such a burden on Nature when, by a little foresight and a perfect reapposition of parts, he could make her task so easy that a final result was certain? One condition he presumed as absolutely essential for success—namely, that the surgeon who applied the "new method" of treatment for fractures of limb bones must be a master of surgical cleanliness.

No one who has followed the history of discovery and invention in any line of human endeavour will be surprised to know that Arbuthnot Lane was not the only man who was seeking for a better method of treating fractures of the leg. Broken legs are as common in the port of Antwerp as in the port of London. We learn from Dr. Albin Lambotte, surgeon to the Stinvenberg Hospital, Antwerp, that his brother, Elie, treated oblique fractures of the tibia in 1888, 1889, 1890 by direct operation, planned on the same lines as those which had appealed independently to Arbuthnot Lane. Each man had met the same problem, and solved it in a similar way. The Belgian surgeon, having taken the first step, appears to have desisted, while the English surgeon, having once begun, kept on until he had awakened the surgical conscience of England. That task took him about sixteen years. The first real sign of awakening was in 1910, when the British Medical Association met in London. A Commission was constituted at the London meeting to investigate and compare the results obtained by treating fractures (1) by manipulation and immobilisation by means of splints; (2) by direct operation; (3) by massage and mobilisation. The verdict which appeared two years later, in 1912, was distinctly in favour of Arbuthnot Lane's operative method. The cases treated by his method were relatively few in number—147, while those treated by the usual methods of immobilisation numbered 2596. The result of treatment depended on the age of the patient. The statistics showed that non-operative means gave the better results in cases where the patients were 15 years, or younger; but as regards adults, statistics favoured the method by operation. I do not think the Commission gave its verdict on the statistical results; it favoured the operative method because of its underlying principles.[1]

What are those principles? The first one—the perfect reapposition of parts—is the enunciation of a simple anatomical truth. Bones, joints, and muscles are so shaped and so fitted together that they work as parts of a perfect mechanism; the alteration of one part necessitates an alteration of all the parts. Perfect restoration permits the old machinery to be used without any alteration. The second principle applied in the operative treatment of fractures is a surgical one; the gap between the broken ends is treated as if it were an incised wound, edges are brought together, not by sutures but by screws; the aim is bony union by first intention. The principle of rest is secured by fixation derived from screws and plates in place of outward splints and bandages. The essential feature of Arbuthnot Lane's treatment is the restoration of parts to their original relationships. The method implies immaculate cleanliness. In theory the method seems perfect; the surgeon is expected to give Nature every opportunity of effecting a perfect result.

[1] Sir Wm. Arbuthnot Lane's criticism on the report will be found in his *Operative Treatment of Fractures*, 1914. Sir Robert Jones has reviewed the Report, *Brit. Med. Journ.*, 1912, vol. ii. p. 1589.

The difficulties encountered by the surgeon when he seeks to reappose the broken ends of bones are due, according to Arbuthnot Lane, to anatomical changes of the following nature. The sheaths of bone, muscles, vessels, and nerves are made to fit the exact length of the adjoining bone; but when the cellular spaces of these sheaths, loosened and dislocated by the fracture, become distended and swollen by extravasation of blood and by inflammatory exudates, they become shortened and thus present a powerful obstacle in restoring the bone to its original length. Muscles are regarded, on this hypothesis, as merely passive structures, but I cannot conceive muscles, unless they are paralysed, offering a purely passive resistance to the reduction of fractures.

In his system of treatment Arbuthnot Lane gives the employment of exercise and passive movement a place; fractures in the neighbourhood of the wrist, elbow, and knee, he thinks, should be treated by early movement of the adjacent joints. In this he is abandoning much of the teaching of his predecessor Hilton. Further, he does not consider that operative means are applicable to fractures in the regions of the wrist and elbow.

We have been examining a revolutionary movement in British orthopædic surgery which was initiated at Guy's Hospital. We propose now to shift our inquiry to Liverpool, and note the direction in which the principles and practice of Owen Thomas have progressed in the hands of his nephew and successor—Sir Robert Jones.[1] Our task is an easy one, for in 1912, when the Report of the British Medical Association appeared, Sir Robert Jones contributed an article to the *British Medical Journal* (7th Dec. 1912, p. 1589), in which the modern problems of surgery, so far as they relate to fractures, are very clearly discussed. We see at once that, although modified and extended, the broad principles laid down by Owen Thomas still serve the Liverpool School as a basis for the treatment of fractures. The advantage of exact reapposition of the broken ends of a fractured bone, the necessity of

[1] Sir Robert Jones assisted his uncle, Hugh Owen Thomas, for eleven years prior to the death of the latter in 1891.

restoring the true axis of the limb, of obtaining true alignment of fragments of the broken bone, so that joints may preserve their normal movements and muscles their normal lines of contraction, are insisted on, and striven for even more strenuously than in the days of Owen Thomas. Powerful traction is the chief means relied on for reapposition of parts—powerful traction combined with deep anæsthesia and skilled manipulation of the parts. Only those who have watched the procedure of the Liverpool School realise the care, time, and trouble entailed by a truthful reapposition; it has always been a maxim of that school, that the less the surgeon does the more Nature has to do. When a fracture has been set, the limb is immobilised in the various methods devised by Owen Thomas. His knee splint, hip splint, walking caliper, and neck sling or "halter-gauge" still supply the means of giving the injured part the "prolonged, continuous, and enforced rest" and the freedom of circulation which are needed for repair. We observe that a splint, modelled on the knee pattern, has been adopted for treatment of fractures of the arm. There is the same insistence on the personal attention of the surgeon in every stage of the healing of a fracture, and the same instruction as to the care which must be exercised in ascertaining that the apparently healed fracture is fit for use, as in the practice of Owen Thomas. Rest is still the basal principle of the Liverpool School—applied in the form of an immobilisation which eases the muscles, shelters the repairing tissue, and allows the circulation to continue without interruption. There is also the same condemnation of immobilisation by means of plaster bandages.

Owen Thomas, we have seen, was not averse to operation; in a case of irreducible dislocation of the thumb, he opened the joint and levered the proximal phalanx into position. In the case of ununited fracture it was at one time the practice of Owen Thomas to operate on the fractured ends and unite them by pegging, but that practice he ultimately gave up. We see in his successors a tendency to increase the scope of operative treatment for fractures. Sir Robert Jones agrees with Sir W. Arbuthnot Lane that many cases of oblique fracture of the tibia are best treated by operation. In such cases restoration, he believes, may be made easier by cutting the tendo Achillis, and thus eliminating the traction of the muscles of the calf. Having obtained true apposition by operative means, he prefers to maintain the fragments in position by application of suitable splints rather than by the use of screws and plates. Fractures of the patella, fractures of the neck of the radius, and some fractures of the olecranon he would also treat by operative measures. We thus see that the Liverpool and London Schools are in agreement as regards many points in the treatment of fractures; both agree that certain should be treated by operative and certain by non-operative means; they are at one when they insist on anatomical restoration, and on the evils which result from a deformity which interferes with the true working of muscles and joints. Both agree that, at the present time, many fractures are permitted to heal in a manner which constitutes a "disgrace to surgery." Both agree that there is too high a percentage of failures in our modern practice of treating fractures, but they differ as to the cause of the failures. Arbuthnot Lane attributes them to the methods used—attempted reduction by manipulation and attempted immobilisation by means of splints. Robert Jones has replied that it is not the method of immobilisation by splints that is at fault, but the imperfect and unskilled manner in which the method is at present carried out. He instances the case of surgeons who, having imperfectly reduced a fracture dislocation of the ankle, wrap the part in a plaster casing and expect a perfect foot to emerge at the end of six weeks. In such a case the entire cure has been left to Nature. She has been expected to perform the miracle of remodelling the limb. She has to attempt a miracle under the most disadvantageous circumstances, for no plaster bandage can be applied that does not interfere with the circulation of the part. Instead of a miracle there ensues a disaster, because the surgeon has thrust his own task, as well as Nature's, on the reparative powers of the part. In brief, the Liverpool School still maintains that the old means of treating fractures by manipulation and splints is the best for the majority of cases—if such means be skilfully applied. Whether we adopt the practice of the one school or of the other, we must be masters of the technique we would employ. Except in the hands of the most practised and proficient, the method by operation must entail the greater risk to life and limb.

We again meet with a difficulty, which often crosses our path in these inquiries: when should the period of rest end and the period of movement begin in the treatment of injuries of bones and joints? Owen Thomas's instruction was: when pain is gone at the site of healing, and the patient has regained complete voluntary control of the limb or part. He also regarded voluntary movements, however slight, as having a much greater curative effect than passive movements or massage. These principles still guide the Liverpool Orthopædic School. Sir Robert Jones recommends window-cleaning, hammering, planing, treadling, to clear the stiffness from the joints of wounded soldiers in preference to the monotonous passive movements given by special and elaborate machines. We have ample evidence of the success with which the Liverpool School has applied volitional exercises to the mending and improvement of limbs wrecked by infantile paralysis. We also note a tendency in the modern school to disregard Owen Thomas's condemnation of forced movements for the breaking down of adhesions. Fractures at the lower end of the radius are moved and massaged at the end of three weeks; fractures near the elbow are also mobilised at an early date.

The Liverpool methods are based on a knowledge of physiology. We have evidence of that in the care taken to safeguard the circulation of the injured limb and in the abandonment of plaster splints; in the use made of damming or venous congestion to increase the activity of bone-forming cells; in its efforts to secure anchylosis at a functional angle in the cases of ankle, knee, hip, shoulder, elbow, and wrist injuries.

We have visited London and Liverpool with the aim of ascertaining the principles which guide certain modern British surgeons in effecting repair of the mechanical framework of the human body, for all operations and appliances which have that aim in view lie within the scope of orthopædic surgery. I propose now to make a brief survey of another field of British orthopædic surgery—the treatment of deformities of the body both congenital and acquired. As an instance on which we may test our modern knowledge and practice, I shall select two of the commoner deformities which occur at the knee—*genu valgum*, knock-knee or in-knee, and *genu varus*, out-knee or bow-leg. I select these deformities because they supply us with a convenient means of comparing and testing the orthopædic practice of one school with another and of one country with another. The method which a surgeon applies to the treatment of deformities of the knee gives us a critical test of his principles and practice. The static deformities of the knee serve such a purpose very well. There is no doubt that so far as our country is concerned, if we would obtain our information at first hand, we must visit Glasgow and examine the methods used by Sir William Macewen. In the late seventies of last century Scotland was swept by a surgical movement, which, as we shall see later, had its beginning in Germany some thirty years earlier—the subcutaneous method of cutting bones for the rectification of deformities. We have already noted that in 1865 Louis Stromeyer Little performed the first osteotomy in Britain, and that in 1871 Adams introduced his operation for the relief of bony anchylosis at the hip-joint. In 1875, my former teacher, Sir Alexander Ogston, devised what I regard as one of the neatest and most ingenious operations ever introduced into surgery. He proposed to restore the transverse axis of the knee-joint in cases of genu valgum, and bring the leg into the supporting axis of the body, by a simple oblique section of the internal condylar part of the lower extremity of the femur. The operation, excellent in design, proved a disappointment in practice. In the same year, Macewen, then in his twenty-seventh year and on the staff of the Glasgow Royal Infirmary, also sought for an effective means of rectifying knock-knee. Two years later, in 1877, he evolved the method now adopted in every country. He restored the transverse axis of the joint, by a simple operation—a subcutaneous, transverse section of the femur in its supra-condylar region.

In 1880 he published a small book under the title of *Osteotomy*, a little classic where we learn the steps by which he successfully solved the operative treatment of knock-knee. By his operation he sought to obtain the same results as followed the successful treatment of a fracture—a perfect and anatomical restoration of parts of bones, joints, and muscles. The casual reader is apt to think that Macewen had solved the problem of knock-knee; he has never claimed to have done more than offer a palliative—the best available. We are apt to think that all the problems relating to the causation and treatment of knock-knee have been solved. In reality that is not so. The true criterion for measuring the standard of surgery in any country is not the success in the application of operative treatment, but the success which has been attained in its prevention. But before means for prevention can be adopted, we must have discovered the cause of a disease or deformity; nor can our treatment be called rational or scientific until the cause is known. Once the cause is ascertained, we know for certain that we have also become masters of the means of prevention. I may appear to labour this point, but I am persuaded, if our modern British School of Surgery has a defect, it is the tendency to remain satisfied with success in operation. We must do more; we must dig beneath the surface of current knowledge and try to discover the true cause of the disease or deformity.

If Sir William Macewen has not explained how the condition of knock-knee has been brought about, he has done the next best thing—he has put us on the road to discover it. In 1880 the cause which found the widest acceptation was the one which ascribed the outward bending of the leg (for abduction of the sub-genual part of the limb is the true nature of knock-knee) as due to a spastic or chronic contraction of the biceps and ilio-tibial muscles. Macewen rejected that theory because he observed that the softening of the bones which accompany the production of knock-knee in children and adolescents, or of the opposite condition — out-knee — was accompanied by a slackness, a want of tone, in the muscles of the limb. He therefore argued that the condition of knock-knee was not due to any abnormal contraction in one set of muscles. If such a deformity were due to muscular action, we ought to see the same influence at work in the upper extremity; that we do not see; such deformities are confined to the bones of the lower extremity.

If we turn to the last edition (1912) of Mr. Tubby's excellent text-book on Orthopædic Surgery, we find that muscles are still absolved from any causative relation to knock-knee or to out-knee, yet the thesis I am to maintain, and which I hope to prove to the reader's satisfaction, is that all static deformities of the knee are due to a definite and definable lesion in the action of the muscles of the thigh.

There is a tendency nowadays to be sceptical of what our forefathers were wont to describe as "laws" or "general principles." That scepticism is only justified if we fail to test such laws by accurate and continued observation. The first principle which I wish you to consider is one I have named the "law of ligaments." No ligament is ever used for the continuous support of any joint or part; that is true of the human hip and the ox's neck. Place your fingers on the neck of the browsing ox or in front of the hip of the standing man, and you will find the muscles are in a state of contraction; they are saving the ligaments from strain. Let us see how such a law helps us to explain the static deformities of the knee-joint.

I want to fix your attention on a living model, because it is in the living body—not in the dead one—we are most likely to obtain assistance. The model should be nude, looked at in profile, with his left leg carried forward—the first movement of a step. We now watch his right limb swing forward as the body is supported on the left limb; as the right leg swings forward we note that the knee bends. In that movement the muscles which flex the knee—the hamstring group—act *from* the pelvis and on the upper end of the tibia and fibula. But in the next moment, when the weight of the body is poised on the right thigh and the left leg swings forward, the action of this group of muscles is at once reversed. The upper end of the tibia and fibula becomes the base from which the hamstrings act; the pelvic girdle forms the system of levers by which these muscles balance the body on the knee. That these muscles are in action whether we stand on one or on both legs, there cannot be a shadow of doubt; one has only to place a finger on them to be convinced that all of them are tense when we stand.

Let us look at these thigh muscles more closely. There are six of them. Four of them—semi-membranosus, semi-tendinosus, gracilis, and sartorius—spring from the inner side of the knee—they form its inner muscular supports; while two of them—the biceps and tensor fascia femoris—spring from the outer side. Of the outer muscular supports, the biceps acts on the ischial extremity of the pelvic (hip-bone) lever; the tensor acts on the iliac extremity of the pelvic lever. The pelvic lever, made up of ischial and iliac arms, is poised on the femoral head at the junction of its ischial and iliac parts. Of the inner muscular supports, the sartorius acts on the iliac end of the pelvic lever, the biceps and semi-tendinosus on the ischial end, while the gracilis acts on the transverse or pubic arm of the pelvic lever. As long as we stand, these muscles are continuously in action balancing the body on the upper end of the tibia; the fixed point from which they act is the upper extremity of the tibia; they come into action with every step we take. In reality the proximal ends of the tibia and fibula are bound to the pelvic girdle by great muscular straps, exerting a continuous pull so long as we are on our feet, standing or walking. We can stand only so long as these muscular straps are in action. If those of the inner side give way, the strain falls on the internal lateral ligament; the ligament lengthens, as every unsupported ligament must do when it is subjected to a continuous strain; so do the parts of the femur and tibia to which the ligament is attached. A small and soon a larger gap appears between the articular surfaces of the internal condyle and the inner tuberosity of the tibia. The condition of knock-knee is thus developed, because the abnormal strain on both the internal and external condylar parts of the femur alters the rate of growth in the outer and inner parts of the epiphyseal lines. If, on the other hand, the outer muscular supports give way, the condition of out-knee results.

What are the conditions which lead to a functional breakdown of the supporting muscles of the knee? We have already seen that these muscles are called into action and continued in action by a reflex-arc system. Why should this system break down in conditions that lead to softening and disturbed growth of bone? In rickets the muscles are flabby, they have lost their reflex tone. That may be due to a tonic substance which acts directly on the reflex-arc system. But there is a simpler explanation. The reflex tone of a muscle depends on its state of tension—a state depending on the absolute rigidity of the bones on which they act. If the femur becomes softened and loses its rigidity, the reflex system which regulates the static muscles of the thigh must be affected. The primary cause of static deformities of the knee lies in a defect or change in the osseus system, but the structures which are immediately to blame for the deformity are the muscles which I have named above. In a rickety child the ligaments of the knee-joint have thrown on to them a static muscular function, with the result that deformities are produced. If we would prevent such deformities, it is clear that we must relieve the lower extremities of their static function, we must prevent the muscles from coming into use, until the skeletal parts have regained their normal health and rigidity; so soon as the normal degree of rigidity is regained, the muscles will resume their normal static, reflex function.

LEWIS A. SAYRE,
BORN 1820; DIED 1900.

CHAPTER XI

THE FORERUNNER OF THE MODERN AMERICAN SCHOOL OF ORTHOPÆDIC SURGEONS

In the preceding chapter cursory visits were made to London, Liverpool, and Glasgow to note the principles and practice of certain modern schools of British orthopædic surgeons. In this chapter a rapid tour is made along the eastern sea-board of the United States to note the rise and expansion of orthopædic surgery in America. We shall find the principles and practice of John Hunter established in Philadelphia by a favourite pupil of the master's later years—Philip Syng Physick. Then came the Stromeyerian period and tenotomy in the thirties of last century. We have seen how Stromeyer's methods were carried to England in the commencement of 1838 by W. J. Little. In the previous year (1837) they were carried to America by William Ditmold, who was born in Hanover in 1808, being four years junior to his fellow-townsman, Stromeyer. Ditmold settled in New York, was appointed to Bellevue Hospital in 1841, dying in the city of his adoption in 1894 at the age of 86. The real founder of the American School of Orthopædics was Louis Bauer, also a pupil of Stromeyer's, who emigrated from Germany and settled in Brooklyn in 1852 when he was in his thirty-eighth year. He afterwards moved to St. Louis, but it was during his

Brooklyn period that he saw the rise of the New York school led by Lewis A. Sayre, Henry G. Davis, and Charles F. Taylor. A glimpse at the work done by Bigelow and afterwards by Bradford brings this chapter to a close.

IN our search for the best means of restoring normal movements to soldiers who have been maimed by wounds, or suffered a physical breakdown under the stress of military service, I propose in this chapter to pay a brief and cursory visit to the United States of North America. Clearly the space is too limited to permit a survey of the great and extensive surgical centres which lie between the Atlantic and Pacific sea-boards; indeed I propose to devote the greater part of the space to New York itself. Nor would it be possible to follow the methods of treatment applied to every kind of deformity or disability. It will serve our purpose best to restrict our attention to one part of the body, and there are many reasons why we should select the spinal column. The treatment of spinal diseases, injuries, and deformities has engaged the attention of American surgeons from an early date; there is no part of the body which more directly tests and taxes the anatomical and physiological knowledge of the orthopædic surgeon.

Before we set out, it will be well to review our knowledge of the essential points relating to the mechanism of the living spine. When we look at the naked back of a man sitting or standing in front of us, we are really looking at one of the most wonderful acrobatic feats to be seen in the animal kingdom. In such an attitude twenty-four segments or vertebræ are being delicately and steadily balanced, one upon the other, by means of the elaborate system of levers and muscles with which each vertebra is furnished. The inter-vertebral disc on which each vertebra is poised is at once one of Nature's oldest and most ingenious contrivances. It is a water-cushion surrounded by a strong, but loose ligamentous covering. Only when the spine is bent or rotated do these ligaments, which enclose the inter-vertebral cushion, become stretched, the fibres on the convex side of the curve become taut and strained, while those on the opposite side are flaccid and bent outwards. The lowest vertebra of this flexible, weight-supporting spinal rod is based on the sacral or pelvic cushion; on the highest is poised the head.

The leverage and motor system of an individual vertebral segment is best seen in a vertebra taken from the dorsal region. The posterior or spinous lever bifurcates at its root so as to enclose and protect the spinal cord; an elaborate system of muscles is attached to the spinous lever whereby the vertebral body may be maintained in its correct antero-posterior plane. On each side of the vertebra is placed a lateral lever, rendered enormously powerful by being prolonged as a rib. These lateral levers are also furnished with elaborate sets of balancing muscles. In the living upright spine every one of these muscles—we may allow an average of eight to each vertebral segment, making about two hundred in all—is called into action, not by any conscious effort of the will, but by that reflex mechanism which automatically governs all the static actions of the human body. We have to remember Beevor's dictum that the opponent or antagonist of the spinal muscles is gravity. So long as the muscles balance their vertebral loads truly they have an easy time; but when gravity comes into play and bends the body to one side, then the opponents of gravity, situated on the opposite side of the body, have to come into action. By reason of certain but very imperfectly known conditions, the spinal muscles may tire in maintaining an even balance of the vertebræ. They are then in the position of a draught horse exhausted by an uphill burden; the waggoner gives his beast rest by blocking or jamming a wheel. That is exactly how the muscles of the spine seek for rest when they tire in maintaining the spine erect; they allow the vertebræ to slew round until the movement is checked. The spokes which are employed to jam the vertebral wheel are the ribs. Once the spinal muscles learn the trick of throwing their burdens on the ribs, they are apt to resort to it time after time until what was a functional posture becomes a confirmed deformity. If we wish to rest the spinal muscles or relieve the spinal column of its weight-bearing function, it is clear we can accomplish those ends by keeping a patient extended on his back in bed.

There is one other point, too, we have to keep in mind—the peculiarity of the rotatory movements of the vertebræ in the loins. When we twist our trunk, either to right or left, as when trying to look over a shoulder, it is the spinous levers which swing in the neck and back, the centre of the vertebral movement being on the bodies, but it is exactly the opposite in the loins. In lumbar vertebræ the swing of the spinous lever is prevented by the interlocking of the articular processes; but the intervertebral discs are deep and permit the bodies to swing. The knowledge of that fact is the key to scoliosis.

We are now in a position to proceed with our survey, and I propose to begin in the city of Philadelphia in the year 1826. I have a particular reason for choosing that year and place. Dr. John Rhea Barton, surgeon to the Pennsylvania Hospital and member of a widely-spread Anglo-American medical family, had then performed a very daring new operation on a case of anchylosis of the hip-joint. He wished to restore movement at the hip, and for that purpose had exposed and cut through the neck of the femur by means of an incision made along the trochanteric region of the thigh. Some years later he sought to restore movement to anchylosed knee-joints by an operation carried out on similar lines—free exposure and section of the bones. I have directed your attention to Barton's work to show you that at the beginning of the nineteenth century there was already a native school of American surgery in which new operations were initiated and successfully carried out. Twenty years later (1852) German surgeons introduced osteotomy as a subcutaneous operation.

If we would understand how it was that Rhea Barton came to adopt these new and heroic methods in dealing with anchylosed joints in the Pennsylvania Hospital in 1826, we have to fix our attention on the senior member of its staff—Philip Syng Physick—who in 1826 was a man of 58, and recognised as one of the most daring and thoughtful surgeons in America. In 1788, when he was 20 years of age and had completed his medical apprenticeship under Dr. Kuhn—a pupil of Linnæus—young Physick, the son of a man born in England, embarked for London, and in a grey, foggy day of November 1788 sought out Leicester Fields and became an inmate of John Hunter's great establishment, and very soon a favourite pupil and ardent disciple of the master of British surgery. He helped Everard Home, Hunter's assistant and brother-in-law, to trephine a sheep for "the staggers," and saw a hydatid removed from the brain. On 1st January 1790, Hunter chose him as his house surgeon in St. George's Hospital, and for twelve months Physick worked on the most intimate terms with Hunter, who was then ailing and ageing fast, but as zealous as ever in his favourite pursuits, and particularly interested in all that concerns the mending of broken bones. Hunter tempted the brilliant Philadelphian to make London his permanent home, but that was not in the programme which Physick had marked out for himself. So, spending a year in Edinburgh to become a graduate of the famous university of that city, he returned in the autumn of 1792, a young man of 24, to commence the practice of his profession in Philadelphia. Two years later—in 1794—he was elected to the staff of the Pennsylvania Hospital; in 1805 he became the first Professor of Surgery in the University of Pennsylvania, and among his pupils was Rhea Barton. We do not need Physick's own statement to know that the principles which he applied were those he had learned in London. We have only to look at his practice and see him adopt rest as the chief means of treating fractures; he designed his splints so that they would secure this end—particularly in disease of the hip-joint. We see him apply Hunter's "principle of stimulation" in cases of delayed union of fractures. It is characteristic of the man that he adopted a form of stimulation of his own. There is his famous case of the sailor who had sustained fracture of the shaft of the humerus at sea, and came for treatment eighteen months later with the broken ends still apart. Physick passed a seton through the arm so that it lay between the broken ends of the humerus. In five months he secured a perfect union. It was in 1816, when he gave up the chair of Surgery to fill that of Anatomy, that he made his most important discovery—the absorbability of ligatures prepared from animal tissues. He died in 1837 in his sixty-ninth year. Thus in Philip Syng Physick we have not only the "Father of American Surgery," but also the most interesting and important link that binds together the medical histories of Britain and America.

We are now to move northwards to the city of New York, because it is there we can best get to know the principles which have been followed by American surgeons—particularly by the pioneers of orthopædic surgery. In 1837—the year in which Philip Physick died in Philadelphia—New York, like London, received its Stromeyerian missionary; in London, the missionary was William John Little; in New York he was named William Ditmold. Ditmold was born in Hanover in 1808, and was thus in his twenty-ninth year when he settled in his adopted city to introduce the orthopædic practice of his master, Stromeyer. He became surgeon to Bellevue Hospital in 1841, dying in 1894 at the age of 86.

Ditmold interests us now only in so far as he served to introduce to American surgeons of his time the most approved orthopædic methods of Germany. But our interest is of a very different kind in the case of Louis Bauer, who was also born in Germany in 1814, studied under Stromeyer, and afterwards emigrated to New York and settled as an orthopædic surgeon in Brooklyn. I do not know the exact year of his emigration, but apparently it was 1852, when he was 38 years of age. In that year he spent some time in London, watching the practice of Little, learning English, and ultimately becoming a member of the College of Surgeons. He became attached to the Medical and Surgical Institution of Brooklyn, where he gave lectures on orthopædic surgery, and at the same time kept up a running fire of criticism on the practice of the American leaders of the New York School of Orthopædic Surgery—Charles Fayette Taylor, Henry G. Davis, and Lewis Sayre. It will repay us presently to examine the nature of his criticism. Thomas of Liverpool also kept his eye on Louis Bauer, and was sometimes particularly severe in the criticism he measured out to him; his sharpest arrows, however, he reserved for

Bauer's rivals—Davis and Sayre. "In my early life," Thomas wrote, "I was in my practice a close imitator and an ardent admirer of my friend, Dr. Louis Bauer—the best exponent of American orthopædics." Everything we can now learn of this American pioneer justifies the acumen of Thomas's earlier judgment. In 1864 Bauer published his lectures on orthopædic surgery in book form; a second edition appearing in 1868. I daresay many, like myself, have picked up these lectures, expecting to find them designed for an ephemeral, perhaps selfish, purpose, and have been surprised to find them not only excellently written, but exceedingly instructive. Bauer's orthopædic lectures, although they may not mark the beginning of a new movement, yet certainly do represent a real step in our progress, and deserve a place on the bookshelves of every surgeon who desires to apply treatment on a rational basis. The curse of orthopædic practice, in Louis Bauer's opinion, was that form of empiricism which we call quackery. In his opinion there was only one cure for the men who were under its spell: a knowledge of the mechanism of the living human body. We have only to open his pages to see that he had mastered all that Marshall Hall, Delpech, and Duchenne could teach him concerning the action of muscles, nerves, and reflex centres. He had extended the knowledge so gathered by personal observation, and applied it with success to the treatment of deformities which are produced by irregular, inco-ordinated muscular action. He was an unflinching advocate of the beneficial action of rest. The surgeon, he maintained, had no better means of securing repair for an injured or diseased joint than by resting it. As is ever the case with those who give that doctrine an unconditional observance, there came to him, particularly in the earlier years of his career, moments of rude awakening. When he had tided joint cases successfully through their acute and critical phases, and only stiffness remained to be cleared up, bone-setters and sprain-rubbers would step in and, in a night, reap the harvest he had sown. From such unlicensed practitioners Bauer became convinced that "Action" as well as "Rest" must have a place in the surgeon's armamentarium.

He maintained that the various methods of traction which surgeons applied to limbs for the relief of joint disease—particularly those devised by his New York contemporary, Henry G. Davis—were merely clumsy methods of securing rest. It is true, as he asserted, that traction cannot separate the opposed surfaces of an inflamed joint; but he quite overlooked the peculiar virtue of the method of traction applied by Davis to the lower extremities in 1860. Davis in New York and Thomas in Liverpool had independently worked out the same idea—the idea which finds its most practical expression in Thomas's knee splint. That splint, as we have seen, is designed to serve as a temporary skeleton for the limb; when it is fitted in place with its ring resting on the natural base of the lower limb, the pelvis, and the muscles of the limb yoked to its foot-piece by a traction apparatus, then the splint serves exactly the purpose of a temporary skeleton. The tonus and contraction of the muscles, instead of being conveyed through the diseased bones and joints, are conducted by the framework of the splint-skeleton. If Bauer failed to see the merits and rational design of Davis's splint—Sayre was quick to see them—he at least saw that the method of traction which yoked a movable patient to the foot of an immovable bed was based on a foolish principle.

As regards the treatment of caries of the spine, we again find Bauer insisting on rest as the beginning and end of treatment. There was only one way—rest in bed in the supine position. The need for fresh air, sunshine, and movement he regarded as vulgar delusions. Spinal supports of all kinds—corsets, mechanical appliances, particularly the "spinal assistant" of Charles Fayette Taylor—were pure "make-believe," because they did not and could not relieve the spinal column of its weight-bearing function. He declared, quite truly, that no instrument had been invented, or could be invented, which would serve to transmit the weight of the head and shoulders and trunk to the pelvis, and thus relieve the spinal column of the burden which falls on it when a patient stands.

We should make a mistake were we to accept Bauer's estimate of his New York contemporary, C. F. Taylor.

His was a very distinctly marked character — recalling in many points our English surgeon, H. O. Thomas. He was born in the same year, 1834; he, too, found that he must design and forge his own appliances if his ideas were to be carried out successfully in practice; he also knew that the effective application of mechanical apparatus entailed endless demands on the surgeon's time, skill, and care; he drew his knowledge, not from books, but from his personal experience. He was a medical graduate of New York, and in 1857, when 23 years of age, settled down there to orthopædic practice. From the very beginning he devoted himself to the treatment of caries of the spine — particularly incipient cases. The disease at that time had a particularly evil reputation—one which Taylor regarded as more than justified after he had applied the forms of treatment which were then taught and used. He laboured sixteen years in perfecting an appliance—his "spinal assistant"—to act as a temporary spinal column. He chose the pelvis as the base on which to rear his artificial spinal support, making the pelvic fit perfect. On the pelvic base he erected a pair of vertical steel supports, to which he tried to transfer the weight of the upper part of the body by shoulder supports. It is not his being a pioneer in the elaboration of spinal supports that has induced me to bring him to your notice, but because I am under the belief that his example in playing the part of an expert mechanician has had a very permanent influence on the modern American School of Orthopædic Surgery. In 1866, when he was 32 years of age, he founded the New York Orthopædic Dispensary. We see the degree of importance he attached to the proper make and fit of mechanical apparatus when he gave as much space and care to the rooms in which appliances were to be forged and moulded as to the wards set aside for the treatment of patients. It was apparently Taylor, more than anyone else, who gave mechanical appliances their important place in the scheme of orthopædic surgery in America.

We come now to a big man—one who altogether captures our interest—Lewis A. Sayre, born in Madison City, New Jersey, in 1820, but for all that a true son and citizen of New

York. He sailed the open sea of general surgery; the inland waters of a speciality were too narrow for him. Nature had given the ship he sailed ample lines and a heavy armament, he could and did use her for fighting purposes; he would send a warning shot across the bows of an enemy craft in the best of good-humour, but for the greater part he used this ship of his not for his own ends either in fighting or in profiteering, but for rescue purposes. He was the first, so it is said, to quarantine a cholera ship; I am sure the storm of opposition which his prompt and daring action aroused in New York never cost him a minute's sleep.

We are to make Sayre's acquaintance on a November afternoon in 1874, in his Out-patient Department attending to poor and ailing children drawn from slum districts of New York. He was then 55 years of age; he had seen thirty years of similar service in the Charity Hospital in Blackwell's Island in the East River; he had been already twenty-one years on the staff of Bellevue Hospital; thirteen years before (1861) he had been appointed to the chair of Orthopædics, then newly established in connection with that hospital. At the very beginning of his career he had served an apprenticeship in anatomy. At the time we meet him his knowledge and experience are ripe. His eye fell on a type of case with which he was only too familiar—a slum-child carrying its body stiffly and guardedly. For many years he had been searching for a means by which his out-patient dispensary could give effective help to poor children suffering from early condition of caries of the spine. That afternoon an idea occurred to him. He had been in the habit of applying to the back of such cases a moulded plaster of Paris splint, but he now resolved to see how a splint which enclosed the whole trunk would answer. He took the suffering boy and suspended him from a tripod by slings applied to the head and shoulders to straighten the spinal column and relieve it of its weight-bearing function. Suspension gave ease, and he proceeded to fix the boy in the suspended attitude by encircling the trunk in a plaster of Paris jacket. He feared as he enclosed the thorax he might produce suffocation, but his fears proved groundless. He did forget, however, that the stomach needed room at meal times; that he made amends for on subsequent occasions by placing a pad on the epigastrium before moulding the plaster jacket. He laid the boy on a sofa to dry, and found, when the plaster was set, that he was not only relieved of pain, but that he could use his limbs with greater freedom and security. His mother was ordered to bring him back in ten days; but weeks elapsed without his reappearance. About three months later Sayre captured his little patient in the slums, and carried him to his lecture theatre to show his students the result of his new method of treatment. The plaster jacket, he found, still served its spinal purpose satisfactorily; it had not effected a cure, for when it was undone and removed, the boy was unable to sit or stand in comfort. The jacket was once more reapplied and the boy sent home again.

Further experience convinced Sayre that the method he had invented was not only cheap and easy of application, but effective in combating and curing one of the most disastrous blights of youth. He was philanthropist and missionary as well as surgeon; he started a crusade and succeeded in awakening the world to the value of his discovery; many a mother and many a surgeon, it is said, has blessed the name of Lewis Sayre. But time has proved that Louis Bauer was right when he declared that there were no means of giving the spinal column rest except by lying down. Experience has shown that all forms of ensheathing plaster splints have certain grave disadvantages. The circulation of a limb or part of the human body which is encased in plaster of Paris is reduced to exactly the same state as the circulation within the cranial cavity or within the medullary cavity of a bone; each pulse beat has to drive out from the veins the equivalent of blood entering by the arteries; the whole burden of the circulation in the immobilised limb is thrown directly on the heart and therefore the circulation of the diseased part suffers. That is relatively a small matter, but there is a greater danger. One knows how slight a cause will give rise to the circulatory condition we call compression of the brain. A plaster splint, applied sufficiently accurately to a limb to serve the purpose of immobilisation, must press on certain capillary areas of that limb —a damaged limb. We have therefore the risk of areas deprived of blood and nourishment, areas of ischæmia and destruction of muscles. The application of a plaster jacket or splint relieves the surgeon and the parent of a daily burden of care, but the burden they have relieved themselves of has been cast on the natural reparative powers of the spine or limb —an injured or diseased spine or limb with a damaged circulation.

The strong personality of Lewis Sayre is seen in his *Lectures on Orthopædic Surgery*, first published in 1876. I cannot conceive that these lectures will cease to be read by surgeons; every chapter is a sheaf from his own harvest of experience. We see in him, as in his present-day successors, a distrust of general principles; methods of treatment, he thought, must be framed to meet the needs of each individual case. In disease of the hip-joint, although he agreed that it was necessary to ensure rest in the early and acuter phases of the disease, he also maintained that such rest damaged the muscles and other structures which surround the joint, and that movement should be given to the joint at the earliest possible moment. Sprains he treated by rubbing, compression, and movement. He had arrived at a rational explanation of knock-knee; it was due primarily to a failure of the muscles which supported the inner side of the joint. His observations on muscles of the limbs maimed by infantile paralysis are of immediate and practical importance. He observed that some muscles which were believed to be completely paralysed and had undergone fatty degeneration could give a slight and evanescent response to galvanic stimulation many years after the onset of the original disease. He saw that such muscles, after having been stimulated, required a long rest before they would respond again to another galvanic stimulus. He was of opinion that the stimulation of paralysed muscles by electrical means, particularly in the earlier stages of that condition, did the muscles an injury. He laid aside the mechanical appliances he had been using for the treatment of paralysed limbs and devoted himself to the nursing and education of individual muscles. He had the observant eye, and was quick to realise the significance of a clinical fact and always ready to modify his treatment to fit his more ripened knowledge.

In our research for guidance in the treatment of disabled soldiers we have been examining the orthopædic practice of certain New York surgeons during the latter half of the nineteenth century. We may continue our search by moving northwards along the coast to Boston, the scene of Bigelow's labours. His treatise on the hip-joint is a standard example of how a knowledge of anatomy can serve the needs of surgery. Our visit to Boston, however, has another object in view. A survey of modern orthopædic practice—even were I fit to undertake such a task—would take us too far afield. Anyone wishing to ascertain the principles which guide and regulate the orthopædic practice of modern surgeons may well find them in the last edition (1916) of *Orthopædic Surgery*, by Dr. E. H. Bradford, Emeritus Professor, and Dr. R. W. Lovett, Professor of Orthopædics in Harvard University. Professor Bradford commenced his career with the capitalised experience of Bauer, Taylor, Davis, Sayre, and Thomas at his disposal, and he, I think, is the chief link between the old school and the new one. There is no school in the world so active in the pursuit of new and improved means of treatment, so little satisfied with those at present at its disposal, as the modern school of American orthopædic surgeons. It seeks help from every line of new endeavour in anatomy, physiology, pathology, physics, electricity, and particularly in mechanical appliances. It inclines to draw its inspiration from recent discoveries made abroad rather than from the solid experience of its predecessors of the nineteenth century. One has only to turn to the recent monograph on the *Treatment of Infantile Paralysis*, by Professor Lovett, to see how thoroughly modern physiology can be applied in the restoration of muscular function. Our main aim, however, is to ascertain how the modern American surgeon faces and solves the most difficult orthopædic problem presented by the human body—spinal deformities due to static causes. This problem is one of immediate interest to the recruiting surgeon. How are we to prevent such blemishes and how cure them once they have become hardened deformities? All are agreed that a true knowledge of their production will provide us with a certain means of preventing their occurrence and the most likely means of removing them if they are once established; all are agreed that such deformities are, in their earlier stages, purely temporary in character, disappearing completely, or almost completely, by voluntary effort. All are agreed that the earlier stages can be remedied by proper attention to the work of the spinal column—by seeing that its musculature has proper periods of rest and exercise. All agree that a lazy habit of mind on the part of a boy or girl may have some share, but not an essential share, in producing the deformity. The static muscles of the spine are mainly regulated by a reflex mechanism centred in the cord; this mechanism may break down from several causes. Apparently its efficiency depends on the full action of the glands or organs of internal secretion, and a change does come over that system at puberty. As yet we can only suspect, not prove, that some disturbance of this kind is the immediate cause of a loss of static function in the spinal musculature.

When, however, the deformity becomes fixed, the bones permanently moulded into new forms, and the ligaments contracted or lengthened to fit new conditions—what is to be done? As in heart disease, perfect compensation may be attained by an increased strength and action on the part of the spinal musculature, and the progress of the deformity may thus be arrested, or compensation may fail and the deformity become progressively worse. I think I am right in saying that the prevailing opinion in the United States to-day is that such deformities should be obliterated and normal curvatures restored by means of direct and forcible compression. One of the most ingenious methods applied is that elaborated by Dr. E. G. Abbott, of Portland (Maine). He proceeds on the principle that lateral curvatures must be over-corrected—that an abnormal curvature to the left must be compressed until it becomes an abnormal curvature to the right. He believes that the rectification can be most easily effected when the spine is slightly flexed and flaccid. He therefore encases the trunk in plaster of Paris, while the patient occupies a hammock, lying in the supine position. The spine is thus fixed in a flexed posture. The trunk being so encased, pressure is applied by inserting graduated pads of felt between the rigid plaster encasement and the convexities of the curvature. Such pads will exercise a pressure on the convexities of the curvature. Windows are cut in the encasement over the concavities of the curvatures on the opposite side of the body, thus permitting such parts to move towards a rectified position. As the convexities yield, new pads are inserted to produce further pressure and a further degree of rectification. It is important to note that the compressing force is mainly applied to the spine through the ribs; the shoulder muscles are also utilised indirectly.

The matter is worthy of our closest attention, for it involves not only the treatment of static deformities, but the management of every kind of spinal patient. I am sure that Dr. Abbott's methods will appeal to many because of their apparent simplicity and effectiveness. When we come, however, to look more closely into the principles involved we shall find that the method does not rest on a true physiological basis, and that any improvement obtained by Dr. Abbott's method is only apparent and is confined merely to the surface of the body. Sayre was of opinion that the serratus magnus, through its action on the ribs, was somehow concerned in the production of scoliosis; at an earlier date, Stromeyer expressed even more emphatically a somewhat similar opinion. No one who has studied the position assumed by the vertebræ in the earlier and temporary stages of scoliosis, and particularly the position in which vertebræ become fixed in the late and final stages of the deformity, can fail to be convinced of the essential part played by ribs in its production. There are a number of conditions one has to keep in mind in determining the exact share taken by the ribs in producing the spinal deformity to which we apply the designation of scoliosis. The intervertebral cushions are perhaps the most delicately constituted adaptations in the human body; it may be said of them that they have only a moment of prime development which is reached about the nineteenth year. Every decade after that year their delicacy and balance is declining; the suppleness of the spine decreases, and the function of the balancing static muscles is thus eased. The proper levers of the vertebræ—the spinous and transverse processes—are short, and the muscles attached to them act at a disadvantage, as they balance one segment on another. It is otherwise with the ribs; they are long and powerful levers, particularly from the fifth to the tenth inclusive, and the great muscular sheets attached to their lateral and ventral aspects can act at an enormous advantage, compared to the muscles acting on short transverse and spinous processes.

Every one is agreed that the scoliotic posture is primarily assumed to rest the muscles which balance the vertebræ upon each other and maintain the spine in the erect position. It is also agreed that rest is obtained by an interlocking or wheeling vertebral movement, but the part played by the ribs in checking the rotation of the vertebræ and the lateral curvature of the spine has not been sufficiently emphasised. The ribs are the spokes which are thrust into the vertebral wheel and bring rest to the spinal muscles. When the proper spinal muscles begin to yield from exhaustion the vertebræ swing round until they are held up by ligaments which bind their transverse process to the necks of the ribs; the ribs would, if they could, swing round with their vertebræ, but they are held by the powerful abdominal and intercostal musculature. In reality the spinal muscles abandon their static function to the powerful muscles which act on the lateral and ventral parts of the ribs. Thus the vertebræ instead of being balanced on each other come to be braced, whenever the erect posture is assumed, against the dorsal ends of the ribs, which soon begin to yield and become deformed in the manner with which every one is familiar. Every rotatory movement in the dorsal region of a healthy spinal column is compensated by a simultaneous rotation in the lumbar region. We have already noted that the lumbar and dorsal rotations are of a totally different kind. The dorsal is the primary movement; its centre or vertical axis is in front of the spinal cord represented by the vertebral bodies. The lumbar rotation is secondary; its axis or centre lies behind the spinal cord represented by the articular processes. The dorsal rotatory movement is checked by the ribs; the simultaneous and correlated lumbar movement is checked as the intervertebral discs become taut and stretched. The dorsal rotation is the primary one; all our efforts must be directed to prevent its occurrence; or if it has occurred and become fixed, to undo it. If we succeed, the secondary lumbar rotation will become undone.

Scoliosis is a deformity of exactly the same nature as knock-knee and flat foot—one due to muscles becoming unfit for their static purpose and abandoning it to structures which yield under constant pressure and become deformed. All are conditions which should never appear in a really civilised community, because they are preventable. The foot and knee are accessible for the application of direct treatment, but as regards the spine that is not the case. The spine has collapsed within the costal cage, deforming the dorsal end of the ribs to an extreme degree. In the production of the deformity the ribs have been passive agents. How then can we hope to use them—as Abbott seeks to do—as the active agents in undoing spinal deformities? It is true that at an early stage of the deformity the dorsal ends of the ribs might be used as levers to undo the rotation of the vertebræ; it is clear that Dr. Abbott regards them in this way when he applies pressure to the convexity of their dorsal ends to undo abnormal curvatures; but in late stages such pressure could only remove a surface deformity—the deep essential deformity of the spine would still remain, for that cannot be undone by means of pressure applied to the ribs. The true agents of redemption are the muscles of the short levers of the vertebræ—the erector spinæ and its derivatives, which act on the spinous and transverse processes. Any form of exercise, and for preference the exercise obtained by games and other forms of physical recreation, will ensure the full action of these muscles, for the more we use our lungs the more must we use our spinal muscles for respiratory fixation. I will not pursue this matter further. My aim has merely been to show that, when all is said and done, it is anatomy and physiology that are the surgeon's best guides to rational means of spinal treatment.

JUST LUCAS-CHAMPIONNIÈRE,
BORN 1843; DIED 1913.

CHAPTER XII

MOVEMENT AS A MEANS OF TREATMENT

The scene of this chapter is laid in France. If we had taken up the matters discussed in these chapters in their chronological order then this should have formed the first of the series, for France is the cradle of orthopædic surgery. When Hunter was spending his boyhood at Long Calderwood, NICOLAS ANDRY, Professor of Medicine in the University of Paris, published a work to which he gave the name *Orthopedia*. Andry had laid hold of the truth that the living body is shaped by its muscles; that if we would have our children grow up with straight and well-formed bodies we must see that they have proper exercise; that if the body becomes deformed a proper use of muscles will remove the deformity. When *Orthopedia* appeared in 1741, its author was in his eighty-second year; he died in 1742. Andry failed to make the world realise the practical nature of his teaching. Seventy years after his death the Swede, Peter Henry Ling, succeeded. For us the most important figure in France during the Hunterian period is JEAN PIERRE DAVID, Professor of Anatomy and Surgery at Rouen. He was born in 1737 and died in 1784, beginning life nine years after Hunter and finishing nine years before him. For the first time he defined the respective places of "rest" and of "motion" in the treatment of surgical disorders. David was born and bred in

North France; his successor in point of time, JACQUES DELPECH, one of the most brilliant figures which has ever appeared in surgery and the virtual founder of orthopædic surgery, was of the South. He was born in Toulouse (1777), studied at Montpellier, where he became Professor of Surgery and where he was murdered in 1832, at the age of 55. He based his orthopædic practice on the knowledge he had gathered as anatomist, physiologist, and embryologist. The year of his death—1832—marks the commencement of the Stromeyerian school, which came to be represented in France by M. BOUVIER (1799–1877) and JULES LOUIS GUÈRIN (1801–1886). They were, as we have seen, contemporaries of Duchenne. As representatives of general surgery during this period we have selected J. F. MALGAIGNE of Paris (1806–1865) and AMÉDEE B. BONNET of Lyons (1802–1858). Both were apostles of "rest." Then, finally, appears JUST LUCAS-CHAMPIONNIÈRE—a son of Northern France and an uncompromising advocate of movement as a curative agency in the treatment of all conditions which result from injuries and accidents. Championnière was born in 1843 and died in 1913 at the age of 70.

EVERY one of the surgeons whose practice we have followed so far has counselled us to rest a limb if we would aid Nature in effecting a repair of its wound, injury, or disease. The only exception is Lewis Sayre; he permits us to see, both by his precept and example, that there is another agent which Nature may call to her aid in restoring health to a part—namely, Action or Movement. In order that we may ascertain the place which should be assigned to *action* as a curative agent, I propose to lay the scene of the present chapter in France, for we shall find, even in Hunter's time, that French surgeons discussed the respective merits of rest and action in the treatment of injury and disease. Our story opens in the city of Rouen during the year 1778. In that year the Academy of Surgery in Paris had set as the subject of a prize essay: "To explain the Effects of Motion and of Rest, and the Indications according to which either should be prescribed in Surgical Diseases," and Jean Pierre David, a very remarkable man, aged 41, surgeon-in-chief to the hospital in Rouen, and teacher of Anatomy there, was busy arranging his observations and conclusions with a view of capturing the prize thus offered. We may read M. David's conclusions as to the respective places of Rest and Action in the essay as published in Paris in the year 1779, or as translated into English by Mr. Justammond, surgeon to Westminster Hospital in the year 1790.[1] But before looking into the conclusions it will be well to see what kind of man its author was.

M. David, we find, was born at Gex, in the Duchy of Burgundy, in the year 1737—being thus nine years younger than Hunter—and studied first in Lyons and afterwards in Paris, graduating at the age of 24, in 1761. He had been to Leyden to sit at the feet of the great Boerhaave, and although a real student, with an extraordinary capacity for work, had always a weakness for competitions in the writing of prize essays. In the earlier phases of his career we find him devoting his attention to subjects connected with Respiration, Nutrition, and Growth. The manner in which part of a bone became dead, separated, and thrown out was a matter of prolonged inquiry. While still a young man he was appointed Professor of Anatomy and Surgery at Rouen. He was clearly endowed with a tireless energy and an adventurous, inventive turn which was not limited to the bounds of his chosen profession. We find him inventing a machine for driving piles into deep waters; devising a method of causewaying roads; he also submitted plans of a fire-proof ship to take part in the siege of Gibraltar. About the same time that he was engaged on the essay which now merits our attention he was becoming involved in mercantile speculations which ended disastrously in his premature death in 1784, at the age of 47.

When we look into his famous essay, if we except his theory of humours, we find very little that is speculative; all his conclusions are based on the sober and accurate observations of his own cases. In certain diseases of joints, it was his aim, as it was also the aim of his contemporaries, to bring about a cure by an obliteration of the joint and a union of the involved bones. In brief, a diseased articulation was treated as if it were a fracture which had resulted in non-union. While his contemporaries sought to accomplish their aim by irritating the joint, by moving it, or injecting substance into it, David sought the same end by keeping the diseased parts at rest; he observed that his method was the

[1] There is a copy in the library of the Royal Society of Medicine.

more successful. He concluded, therefore, that to bring about anchylosis—to effect the union of bones—rest was beneficial, action was injurious. On the other hand, if anchylosis were to be avoided, as in injuries to the elbow or knee joints, then the opposite practice must be pursued—that of action or movement. In all cases of wounds and injuries rest was the proper treatment in the healing stages; abscesses when opened should have their walls pressed together. Caries of the wrist or of the ankle he treated by applying splints and giving the diseased parts rest. In caries of the spine the patient was placed supine in bed. "Time and rest will cure all joints," he said. "Absolute rest is an imaginary thing," and "Inaction does not produce disease" are other of his sayings. There were conditions which called for motion. The injured or diseased part was ripe for motion when consolidation or resolution had occurred. Natural motion was best; if there was a slight degree of motion, Nature, he held, would soon make it more. He evidently preferred voluntary to passive movements in restoring the action of parts. When there was scurvy, or gout, or chronic rheumatism, then motion and action were the forms of treatment needed for restoration. It was M. David's opinion that Nature had two resources in repair—she might call for rest or for action; each had its proper time and place in the treatment of disease.

If we were to follow strictly the thread of our story and watch the manner in which French surgeons applied David's principles of Rest and Action to the treatment of wounds and injuries we should proceed to Paris and watch the practice of Malgaigne during the fifties and sixties of last century, or, better still, proceed to Lyons and observe the manner in which Bonnet treated diseases of joints. Before proceeding to that point in our narrative there are at least two men who demand our attention. One is Dr. Nicolas Andry, Professor of Medicine in the University of Paris, who, in the year 1741, when he was an old man of 81 and M. David merely a boy of 4, wrote a very remarkable book on the prevention of deformities in children, and gave it the title *Orthopædia.* He coined that word (ὀρθός, straight; παιδίον, child) to designate the branch of medical learning which has to do with the prevention and cure of the physical deformities of children. We have only to glance through the crisply written pages of *Orthopœdia*—pages which were written long before the Hunterian period had dawned in England—to be convinced that Andry had grasped to the full the rôle of muscles as body-moulders. For him, "action" was a more important means of treatment than "rest." For him, the body of a boy or girl was a plastic thing; by properly ordered exercises, deformities could be prevented or cured. In M. Andry we meet the veritable founder of many of our modern orthopædic practices. He was dead and buried seventy years before Henry Ling formulated his series of gymnastic exercises in Sweden.

If the author of *l'Orthopœdia* deserves a passing notice, then M. Jacques Delpech, the author of *l'Orthomorphie* merits our most serious attention. We have come across Delpech's name before; Stromeyer never ceased to acknowledge that it was the operation carried out by Delpech in the hospital of Montpellier in 1816 which suggested to him a subcutaneous section of the tendo Achillis. Delpech has told us why he wished to cut the tendo Achillis; it was to prevent the muscles of the calf producing a deformity of the foot when their opponent or antagonist muscles in front of the leg were paralysed. Paralytic club-foot he knew to be the result of an unbalanced action of muscles. We also know why he sought to carry out the operation by a subcutaneous method. In 1812, at the age of 35, he was appointed Professor of Surgery in the University of Montpellier—the university of his own student days; two years later, the army, withdrawing from Spain, brought hundreds of wounded men to Montpellier. Delpech sought to obtain union by first intention in the treatment of the soldiers' wounds. He was convinced, as many other surgeons had been, that it was the admission of air to wounds which caused them to go wrong. He felt certain that if air could be excluded, then a wound should heal by first intention. It was for this reason he applied a closed or subcutaneous method when he introduced tenotomy as a rational surgical procedure.

Tenotomy, however, was a mere incident in Delpech's busy life. The event which most concerns us is the publication of his greatest work, *l'Orthomorphie*, in 1828–29, when he was just past his fortieth year. One cannot open the pages of *l'Orthomorphie* without feeling that one is in the presence of a Master of Medicine. Indeed, Beclard has declared that Delpech shared with Broussais and Dupuytren a place above all their medical contemporaries, and that in eloquence and in imagination Delpech was superior to either Broussais or Dupuytren. There is no treatise on orthopædic treatment, written during the last century, which stands so little in need of re-editing as the *Orthomorphie* of Delpech. He formulated his methods of treatment after studying the normal actions of the healthy living body. If treatment was to be rational, then it must be based on physiology. He knew the true function of ligaments, they were merely reserve structures which came into action when the muscles became weak and gave way, or when their defence was broken down; he recognised that every muscle in the living healthy body was in a state of tension or tone; in the maintenance of posture each group of muscles was silently balancing an antagonist group. When the muscular balance was upset, then a deformity of the body was developed. He was the real founder of the theory that deformities are due to a lack of balance in the action of muscle groups.

More interesting even than the letterpress of *l'Orthomorphie* is the volume of illustrations with which Delpech supplemented his great work. There we can see the plans and sketches of the orthopædic institution which Delpech first dreamed of, and then finally realised in the park-land country which surrounds Montpellier. His institution was a pastoral symphony where ladies in rational dress disported themselves in sylvan glades, sometimes stretching themselves on hammocks in the sun, at other times swinging from elaborately constructed gymnastic machines set in a garden land.

It was while driving along the country road which took him to the Orthopædic Institution that he met his tragic death in 1832 at the age of 55. He was shot dead by a man who had been his patient, and on whom, it is said, he had performed a radical cure for varicocele. However that may be, there is no doubt that in Delpech's death one of the most outstanding medical careers of the nineteenth century was suddenly brought to an end.

Jacques Delpech has tempted us aside; we must now return to the straight path of our inquiry and ascertain the principles—so far as rest and action are concerned—which regulated the practice of French surgeons in the middle part of the nineteenth century. We shall cite two representatives of that period—Bonnet of Lyons, who published his justly celebrated treatise on the Diseases and Treatment of Joints in 1845, and Malgaigne of Paris, who, from 1831—the year of his graduation—until 1865, the year of his death, made annual additions to the surgical literature of his time. Malgaigne's treatise on orthopædic treatment appeared in 1862. Bonnet and Malgaigne were advocates of "rest." Bonnet laid it down as a principle that to secure rest in any joint it was necessary to immobilise, not only the affected joint, but also the joints which were proximal and distal to the one affected. In the case of disease of the knee-joint, it was not enough, he held, to apply a splint merely to the region of the knee; the splint must be extended so that hip and foot joints were also immobilised if real and complete rest was to be secured. So far as fractures were concerned, Malgaigne advocated reapposition of the broken ends and immobilisation of the limb by splints. He had never seen prolonged rest do harm.

We have seen that on his arrival in Paris in 1842 Duchenne came into intimate and friendly relations with M. Bouvier, the orthopædist. Bouvier was a man of 43 when Duchenne first met him, and had accumulated a wide and varied experience in the treatment of deformities. After graduating in 1818 he applied himself to the teaching of anatomy and physiology, and very early in his career began to specialise in the treatment of deformities—particularly those of the spine. In 1824, while only in his twenty-fifth year, he took the lease of a house and fitted it out as a hospital for the treatment of deformities—particularly those of the spine—which, thanks to the good management of his wife, proved a success. She aided him, too, by purging his manuscripts of their pungent passages; for he had, as is often the case with orthopædic surgeons, a fondness for polemics, and valued priority rather than fees. Bouvier's contemporary and rival, Jules Louis Guèrin, was two years his junior, and is best known for his advocacy of the doctrine that all deformities were due to spasmodic contracture of muscles, and that the rational preventive treatment in cases of curvature of the spine was section of the offending muscles—myotomy. Guèrin had his followers in England and America, but the success of myotomy was shortlived. From the commencement of Guèrin's campaign Bouvier rightly maintained that section was a brutal and irrational method of treating a functional derangement of muscles.

Immobilisation of broken bones and diseased joints had become the accepted practice of French surgeons of the nineteenth century, and the principle of Rest appeared to have found an undisputed place in surgical procedure. The serenity of professional harmony was soon to be upset, however, by the appearance on the surgical stage of Paris of the polite but persistent figure of Just Lucas-Championnière. He needs our close attention, for we shall find him to be a reformer of an uncommon type. He had no wish to bludgeon those who disagreed with him; he knew a better way of attempting reform. He had to tell his senior colleagues that they were doing great harm by treating broken bones, diseased joints, and injured parts by rest; the right treatment was exactly the opposite—a treatment by motion, action, or movement. They almost forgave him because of the manner in which he delivered his message. He was so apparently honest, so courteous and suave in the propagation of a heterodox doctrine, that they could not but listen to him. He had all the gifts of a great advocate—the power of presenting his own views so lucidly and so convincingly that no other view than his seemed tenable or possible. As one is swept along by the arguments in his treatise on the *Treatment of Fractures by Massage and*

Mobilisation, one is tempted to wonder how his predecessors could have been so blind and his contemporaries so stupid. He is not only an able advocate, but has a personality which fascinates. In listening to Lucas-Championnière we must keep not only our ears open, but our judgments fully awake. We want to see how it came about that Lucas-Championnière advocated the treatment of fractures by massage and mobilisation. He was born in the North of France, at St. Leonard (Oise), in 1843, and pursued his medical studies in Paris, taking up his first resident hospital appointment in 1865 when he was 22 years of age. While still a student he became impressed—probably from experience gained in his native district—with the efficacy of rubbing and movement as a means of treating sprains. In 1867, while still a house-surgeon, his attention was drawn to the case of a fractured radius in an old woman of 66; a perfect result had been obtained although no splint had been applied, nor had the woman kept the arm at rest. The young house-surgeon pressed this case on the attention of the Surgical Society of Paris, and set out to treat fractures of the radius with movement in place of rest. He did the same in cases of fracture of the fibula, and believed the results obtained were better than when splints were applied. We must remember that in such cases the undamaged neighbouring ulna or tibia served as perfect splints. We cannot say, therefore, that those early cases were treated without the aid of splints.

In the autumn of 1868, the second of his appointment as internal resident surgeon, he returned from a flying and fortunate visit to Britain an ardent advocate of Lister's discovery. While in Glasgow he met Lister, who was then devising rational means at the infirmary of that city for the prevention of the infection and suppuration of wounds. Lucas-Championnière was one of those born to look to the future rather than to the past for the means of progress; his instant recognition of Lister's discovery shows us how sound his youthful judgment was. He applied Lister's methods to the treatment of compound fractures and had thus opportunities of studying, at the bottom of healthy open wounds, the exact effect of movement and of rest on the various processes involved in bone-repair. He observed that a slight degree of movement at the fractured ends, far from retarding the progress of repair, rather accelerated it. There was a more abundant formation of callus, which he regarded as advantageous. If movements at the site of fracture were too great, or if the fractured ends were jagged and sharp, then the tender, reparative tissues were injured by motion and consequently exposed to infection. Mobility, he said, had to be given in limited doses.

His resident appointments ceased in 1870, when at the age of 27 he took his doctorate. He had to wait four years for his first appointment on the staff of a hospital. As soon as it came, he sought to extend the application of his method of treating fractures. He treated all fractures in which joints were involved by an early application of movement. He became known at the Surgical Society as the "ankylophobe." "Splints," he declared, "render joints stiff and often do them irrecoverable damage." Enforced rest injured cartilage, ligaments, and muscles. It never occurred to him to ascribe the injurious effects to a wrong method in the application of splints. He condemned the principle of rest as bad. On the other hand, he believed that mobility preserved the life and health of muscles, ligaments, and joints; it accelerated the process of repair and shortened the period needed for complete recovery. Thus it came about that Lucas-Championnière gradually laid the whole armamentarium of splints aside, first for one class of fracture and then for another, until ultimately it was only in cases of fracture of the thigh or leg that he applied any means of restraint, and in these exceptional cases the means employed were of the slightest kind.

In 1881 Lucas-Championnière began to introduce another innovation in the treatment of fractures—the application of massage, not only to the parts in the neighbourhood of a fracture, but actually at the site of fracture. "Massage," he said, "was already in its dotage and in its infancy." He claimed that massage allayed almost instantly the pain at the site of fracture; it accelerated the process of repair; it dissipated inflammatory exudates, reducing swelling and tension in the damaged parts; it maintained muscles, nerves, tendons, ligaments, and joints in a state of health. The application of massage to the immediate treatment of fractures and dislocations he counted amongst his chief services to surgery.

Thus we return from our hurried visit to France with some strange problems to resolve. We found Jean Pierre David teaching at Rouen in 1788 that Rest and Motion has each its place in rational treatment. Then came a long period in which rest was regarded as the supreme remedy for all surgical ills. Then a century after David, we find Lucas-Championnière declaring that rest is a dangerous remedy and that "action is life." He held that the surgeon who would help Nature in effecting repair must assist her by supplying regulated and suitable doses of motion to the damaged part. Wherein lies the truth? Some of us can remember Lucas-Championnière when he came to the Royal College of Surgeons in 1912 to pay a tribute to the memory of Lord Lister, and we can understand how the fascination of his personality may have gained acceptance for doctrines which may not prove permanent. It was in the year of his visit to England, and a year before his own lamented death in 1913, that the Commission was instituted by the Council of the British Medical Association to report on the results obtained by the various methods of treating fractures. None of the methods gave perfect results, but those obtained from the method advocated by Lucas-Championnière—massage and mobilisation—were the least satisfactory. If we turn to the semi-official manuals by Leriche, published in 1916 to guide the practice of French army surgeons, we find that the methods recommended are those of immobilisation—those of Bonnet; the sole trace of Lucas-Championnière's influence is to be detected in the earlier mobilisation of articular fractures, particularly those which occur in the neighbourhood of the elbow-joint.

How, then, are we to explain the fact that opposite and apparently irreconcilable methods have been recommended in the treatment of fractures? How are we to reconcile the experience and teaching of Thomas of Liverpool and of Lucas-Championnière of France? The one treats fractures by enforced rest, the other by limited but enforced movement. The first inference we must draw from these facts is that the tissues of the body are endowed with a power of repair so strong that they will effect a cure amidst the most disadvantageous circumstances. Inspiratory movements never leave the fractured rib or broken clavicle at rest for a moment, yet healing is accomplished. Amongst wild animals and primitive peoples we find fractures which have been healed without surgical intervention. Many of them, it is true, would have been better for surgical intervention. Healing will take place whether we mobilise or immobilise the fractured limb. The surgeon who claims success for the means he has applied in treatment is in the same position as Lady Priestley when she claimed that dogs could be rendered immune from distemper by innoculation with vaccine lymph. "Were there control experiments?" asked Pasteur. There were none, and until a surgeon checks his results by "control experiments," he is in the same position as Lady Priestley.

Here, then, in the year 1890 we have two men of wide experience in the treatment of fractures, one in Liverpool and the other in Paris, both eager to assist Nature in the repair of all surgical disorders of the human frame—yet each insisting on giving his assistance in an opposite manner. That fact brings home to us that we have still to pass through years of patient and toilsome observation before we can tell with certainty how far surgical effort of any kind has helped or hindered in securing a perfect result. The process of healing depends on the presence of a multitude of conditions. But to my mind the truth lies more with Thomas of Liverpool than with Lucas-Championnière of Paris.

There are certain matters which do not admit of difference of opinion. No one can doubt that the more exactly the surgeon reapposes the fractured ends, the easier he makes Nature's task; nor do I think that there can be any doubt that repair at the site of fracture is best effected under conditions of rest. Yet it seems to me that Jean David of Rouen had come very near the truth when he said there was a time for rest and a time for motion in the treatment of most surgical conditions. One cannot examine the condition seen in limbs of soldiers, which have been amputated on account of infections following injuries to bones or joints, and in which we find various stages of healing, without being convinced that mobilisation must play a large part in securing the final stages of repair in the treatment of disabilities which follow gunshot injuries of limbs. Muscles are bound to their sheaths by inflammatory exudate, so that their fibres, even if they are healthy, are necessarily limited in their movements. The whole connective-tissue system of the limb has been swept by an inflammatory tide. Round all moving parts, muscles, tendons, and joints lie envelopes of newly organised tissues. Every moving structure becomes tethered. When such a stage of resolution is reached, rest can no longer be of service. I can conceive of no means which will lead to the lessening and loosening of the adhesive bonds and thickenings which surround muscles and joints except those implied under the term mobilisation—massage, passive movements, and active or voluntary movements. Of these three we have many reasons for expecting the best results to follow voluntary movements, however slight these may be. We know that muscles are so set in the limb, so packed together, that their contraction exercises a maximal propelling impulse on the venous and lymphatic circulation of the limb. The voluntary impulse which sets a group of muscles into action exerts, at the same time, a regulating influence on the vaso-motor system of the part. Nor must we overlook the advantages which voluntary movements have over passive movements from a psychological point of view. "If there is a little movement," said Jean Pierre David, "Nature will soon make it a little more." The muscles of the damaged part are the agents by which that increase of motion is to be obtained. The machinery which can set the muscles to work lies in the patient's own nervous system. The only agency which can set the machinery in motion is the patient's own will. The sooner the patient ceases to rely on outside help, and the sooner he comes to realise that progress depends on his own effort, the quicker and better will be the ultimate result. The surgeon must direct, but he can only succeed in having his directions carried out if he has obtained the intelligent co-operation of his patient. He has to seek such forms of exercise which are at once agreeable to the patient and have a bearing on his future needs in life.

CHAPTER XIII

THE INTRODUCTION OF GYMNASTICS AND MASSAGE TO SURGERY

In this chapter we are to search for the reasons which have led surgeons to apply gymnastics and massage in the treatment of injuries, and in the prevention and cure of deformities of the body. For this purpose we first pass to Paris to again consult the aged Nicolas Andry, President of the College of Physicians and coiner of "Orthopædia." He can give us the orthopædic wisdom of the first part of the eighteenth century better than any of his contemporaries. He was born in Lyons, 1658. He studied for the Church, but later took up Medicine, and graduated at Rheims in 1693, in his thirty-fifth year. Then coming to Paris he had a successful career as a fashionable physician. *Orthopædia* contains the experience of a lifetime; it was published in 1741 when its author was in his eighty-third year. Then we return to London to note what young John Shaw, brother-in-law of Sir Charles Bell, has to tell us of gymnastics and massage in the year 1824. John Shaw was one of the products of the Windmill Street School—a school which is without parallel in the annals of Medicine, and could have appeared in no other country save the home of voluntary enterprise—England. The Windmill Street School was run by a colony of London Scots. William Hunter founded the school in Covent Garden in 1746; there his brother John was his pupil, assistant, and partner. Then he built the Windmill Street establishment, to which he moved in 1768, taking the young, but great, Hewson with him as assistant. At William's death in 1783, Cruikshank and Hunter's nephew, Matthew Baillie, kept the school going. James Wilson succeeded Cruikshank, and in 1815, Charles Bell succeeded Baillie. In 1821, John Shaw, at the age of 29, succeeded Wilson. In 1824 he published an Atlas, illustrated with engravings by Landseer, on *The Nature and Treatment of the Distortions to which the Spine and the Bones of the Chest are subject.* In the previous year (1823), and in the year following (1825), he published treatises on *The Nature and Treatment of Distortions of the Spine.* In these works we learn what Shaw had to teach concerning gymnastics and massage. His teaching, we shall find, is worth our attention. He died in 1827 at the age of 35. Having glanced at Shaw's teaching we again pay a visit to the Orthopædic and Gymnastic Institution which Jacques Delpech established in the suburbs of Montpellier in 1825, and of which he gives us a full account in *Orthomorphie*, published in 1827–28. Delpech, the Hunter of France, who was born at Toulouse in 1777 and fell by the hands of an assassin in 1832, at the age of 55, had, as his contemporary, one Pehr Henrik Ling, the creator of Swedish exercises. Ling was born at Ljunga, in the most southern part of Sweden, the son of a country clergyman, in 1776, and was thus Delpech's senior by one year. He studied theology at Upsala, but having the adventurous, roving blood of the Norseman and the spirit of a poet, he became a wanderer in Europe—earning a living as best he could—until his twenty-eighth year, when we find him teaching fencing and presently studying anatomy at the University of Lund, not far from his birthplace. He was then beginning to conceive that he could make his country healthy, happy, and prosperous by exercise. Having seen how Ling came to make his discovery and the manner in which he put it into practice, we return to England to learn how his teaching prospers in our modern hospitals and homes.

IT was the fate of Dr. Nicolas Andry, the founder of Orthopædics, to live in a period when men were convinced that they could improve upon Nature. Nature, they held, had to be tamed and trimmed; the farther they could get away from Nature, the more they had progressed along the path which led to the final goal of civilisation. The human body was no exception; they applied to it the same pruning, training, and shaping as produced their fantastic garden shrubs and bushes. Andry could not conceive that a time would ever come when a mother was not solicitous in securing a properly shaped waist for her daughter—one which was perfectly flat in front, but full and prominent in the region of the buttocks. That was to be obtained by the early and proper use of stays. Within the scope of orthopædics, Andry included not only the fashioning of the waist, but the care of the nails and hands, the shaping of the head and nose, the arching of the eyebrows and the reposition of the ears, and also much of that kind of knowledge which in our day has become the stock-in-trade of the beauty specialist, voice producer, and master of deportment.

Our particular reason for consulting him now is to learn how far he employed exercises or gymnastics and rubbing or massage in shaping the body to the conventions of his time. He had laid hold of one important truth—muscles were the chief instruments in shaping a child's body; it was therefore by playing on these instruments that a physician could accomplish his orthopædic aim. Nowhere does Andry parade any knowledge borrowed from the dissecting room, yet he shows himself an accomplished artist when he comes to play on the muscular machine. He picked up his knowledge of muscles in his professional rounds; he observed that the corpulent man and the pregnant woman walked with a straight back—even an over-extended one, because they were overloaded in front. If a youth stooped, then the natural way to bring the muscles of the spine into action and give him an erect carriage was to make him carry a burden in front of his body, or to hang a weight upon his breast. If the right shoulder was higher than the left, then a burden carried in the left hand would cause the bearer to bend the head and spine to the right, thus raising the left shoulder and depressing the right; or if the spine was not straight and there was a tendency to stoop, then a game in which a light burden, poised on the forehead, had to be carried from one point of the play-room to another would set all the muscles at work which have to keep the spine erect and balance one vertebra upon another. Repeated balancing exercises of this kind, he held, would restore shape and symmetry to the back. If a girl had assumed a "devotional attitude," with sloping shoulders and a scapular region "shaped like the back of a spoon," then the "spinal hunch" was to be pushed and rubbed; in modern language, massage was to be applied to the convexity of the curvature. Exercises were to be given which would lengthen the abdominal muscles, which, as the attitude indicated, had become unduly shortened; the patient was to sleep supine on a hard mattress; cold douches applied to the region of the shoulders would also stimulate the parts and help to make the muscles do their duty. If one leg was apparently shorter than the other, so that the hips and shoulders became of unequal prominence, then the heel of the offending or shortened limb was to be gradually made higher until not only the symmetry of hips and shoulders was restored, but over-corrected and a fault of the opposite kind appeared, for Andry knew that with a "crooked body, as with a crooked stick, you have to over-correct the deformity if you would aim at obtaining the just mean." "Shoes that are too high-heeled," he said, "will make the bodies of children crooked." But if there was a tendency to shortening of the muscles of the calf with a consequent raising of the heel, then he thought walking exercise, with a boot in which the heel had been weighted with lead, was the proper treatment. Whether we agree with him or not in the principle he applied to remedy a contracture of the calf muscles, we must admire the ingenuity and simplicity of his means, and admit that he knew how to apply physiology. His remedial exercises are based on the simple principle that deformities which are due to one form of movement or habit, can be undone by a movement or habit of an opposite kind. We see him apply this principle in the cure of squint; the patient had to practise his exercises in front of a mirror, and he declared that this method yielded better results than could be obtained by the use of spectacles. He knew the evils of posture, and insisted that fine needlework or the printed page of a book should be raised to the eyes—not the eyes lowered to the book. We could not make a greater mistake than to think that all the orthopædic wisdom lies with us moderns, the man who delimited the boundaries of orthopædics in the eighteenth century has still much to teach us.

"Gentle friction," he says in another place,[1] "produces always very good effects upon the blood, whether it facilitates its course from the heart to the extremities or the opposite." He was a believer in massage and also in exercise. "Moderate exercise," he said, "was the best means of preventing and even of curing disease." The kinds of exercise he had in mind were not only the play and games of children, the fencing, dancing, riding, and billiards of adults, but also the toil of ordinary every-day life. He had seen the miracle which hard labour effected when indolent, delicate girls became ward maids in hospitals. We realise he was more than a fashionable physician when he tells us that "there are a great many men who are obliged to labour constantly for their bread, and consequently have it not in their power to submit to all my rules for the pre-

[1] *Orthopædia—The Art of Preventing and Correcting Deformities in Children.* Translated from the French of M. Andry, Professor of Medicine in the Royal College of Physic, Paris, 1743.

servation of health . . . but if, by a turn of fortune, they are able to cease work, then they lose their health." Work and exercise he counted the best medicines for health, but they had to be taken in their proper dosage. "When the muscles began to swell, when breathing becomes difficult, when the skin has become red and sweats, and fatigue is felt," then was the time for the antidote of exercise—rest. The dispute as to the curative value of action and of rest was already ancient in Andry's day. "Asclepiades and Erasistratus," says he, "have boldly condemned all forms of exercise as not only of no advantage, but even prejudicial to health, and recommended rest as the chief preserver of it; but they were very much mistaken in this point. Rest deserves its own share of praise; it is a restorer necessary in the course of a great many diseases." Thus we see that the founder of orthopædics recognised that rest, as well as action, had its own particular therapeutical merit, but of the two, action was that to which he attached the higher value.

Having thus noted the practice of old Nicolas Andry so far as it relates to the use of exercise and massage, we pass from Paris of 1741 to the very heart of London in the year 1823, to learn the reasons which led Dr. John Shaw to apply gymnastics and rubbing to the treatment of the deformities which are so apt to appear in the spines of young and growing girls and boys. The visitor to London can still see the square, plain, red-brick, eighteenth-century building in which he taught and laboured—a building which has many a duplicate in the older parts of Boston and Philadelphia, for William Hunter built the school when the Union Jack was still flying on Washington's flagstaff. The building is within a stone's-throw of Piccadilly Circus, tucked away behind the Lyric Theatre, of which it now forms a part. But in the year 1823 it was known far and wide as the Windmill Street School; Charles Bell, then a man of 49, and engaged on the respiratory function of certain nerves and muscles, was its head; he had his brother-in-law—John Shaw—as his junior partner. At the time we make John Shaw's acquaintance he was 31 years of age, and busily engaged in the

Museum of the school examining a fine and extensive series of specimens which illustrated the various deformities of the spine—for he was already on the surgical staff of Middlesex Hospital, as well as lecturer on Anatomy at the Windmill Street School, and proposed to apply himself particularly to the treatment of spinal deformities. We know how he had become interested in spinal deformities; only three years before, in 1820, he had attended the lecture theatre of the Royal College of Surgeons to listen to his friend and teacher, Dr. James Wilson, then teacher of Anatomy in Windmill Street, deliver Hunterian lectures in which he expressed the conviction that all static deformities of the spine were due to rickets. In the meantime Wilson had died and Shaw had succeeded him on the Windmill Street teaching staff. His study of the Museum specimens, as well as hospital cases of scoliosis, showed him that although rickets was often attended by a curvature of the spine, yet in the majority of cases the bones of the leg and of the pelvis showed no traces of that disease, and he therefore separated the non-rickety spinal curvatures as a distinct class. The accepted teaching at that time, both in England and abroad, was that lateral curvatures of the spine were due to an unbalanced action of the muscles which maintained the spine erect; on one side—the concave side of the curve—the muscles were supposed to be in a state of spasmodic contraction. Such a theory, Shaw found, broke down when he came to examine young growing lads and lasses showing lateral curvature in its earlier stages; there was no muscular spasm to be seen in their backs. Nor did such a theory account for the changes which he found in the spine after death. All the evidence at his disposal, anatomical, physiological, and clinical, led him to suppose that spinal curvatures were due to a weakness in the muscles which balanced the vertebræ when the body was held erect. He saw these curvatures appear in the backs of healthy young people when they were tired by being kept in a standing or sitting posture. The curvatures disappeared as soon as the body again passed into active movement. The appearance of curvatures he regarded as evidence of weakness on the part of the spinal muscles; they were no longer able to carry out their onerous static duties when young people are kept standing or sitting still for a long space of time. Why the spinal muscles had become weak he could not explain, but he observed that they broke down most frequently during spells of active growth.

Having reached this conclusion, we should expect a disciple of the Hunterian school to apply the logical remedy—rest. A tired muscle should recover its strength best if rested. That was the treatment then in vogue in England. Patients with spinal curvatures were laid on their backs—on an inclined plane—for months or even years at a stretch, but the results are said to have been unsatisfactory, at least Shaw came to that conclusion. Another method of giving rest—popular then throughout Europe—was to support the weight of the trunk and head from the pelvis by means of a machine—a sort of spinal assistant—which not only supported, but was believed to actually extend the spine as the patient took his daily exercise clothed in his machine. Shaw condemned both inclined plane and spinal machine on the score that the more the spinal muscles were rested the weaker they became. He claimed support from what he termed the "eternal law of disease." The proper treatment for weak muscles was therefore, in his opinion, not rest, but graduated exercises followed by rest. But there was also another influence at work which set Shaw's mind brooding on the curative effects of action. "We have only to witness," he writes, "the operations of a professional rubber to be convinced that if the practice were superintended by one acquainted with anatomy and pathology, much benefit would result from it, while there would be no risk to the patient's life being endangered by its indiscriminate application." Shaw had seen case after case of lateral curvature of the spine which had been treated for months on the inclined plane under professional advice or encircled by a spinal support, fall into the hands of rubbers and become—or said to have become—marvels of recovery under their treatment. The "eternal law of disuse" combined with the success he saw attend the empirical efforts of rubbers led him to adopt the "principle of action." He used all the tricks of the masseuse—superficial rubbing, shampooing—after the Indian method—so that the deeper structures were pressed; thumping, pinching, percussion, and kneading were all employed. Such was the preliminary treatment applied to the spinal muscles to soften them and make the parts more pliable. Shaw invented a form of friction roller—the first of its kind; then set exercises were given—by means of apparatus of various kinds—apparatus which offered resistance to movements of the trunk with various forms of pulley apparatus; movements of the arms were used to influence the upper part of the spine, and movements of the lower limbs the lumbar part. Between the various doses of spinal friction and exercise were given spells of rest, either by reclining, or by the application of stays or of a spinal machine. Thus his treatment consisted of alternate applications of motion and of rest. In slighter cases of lateral curvature—cases in which all deformity disappears when the patient stoops—he believed that exercises were sufficient to obtain a cure. He found Andry's practice of making the patient walk, keeping a footstool balanced on the head, to be excellent; he realised that it was not the weight of the object so carried that gave the good effect, but the maintenance of the poise or balance. Balancing exercises were best. John Shaw hoped that the time would come when physical exercises would be better appreciated in England. He particularly commended the type of "gymnastic exercises which Mr. Clias of Berne" was then introducing to the English public. We thus see that in the year 1823 one man at least had realised that gymnastics and massage had a place in surgical therapeutics. Unfortunately John Shaw's influence disappeared soon after the publication of his writings on the spine. He died from a "slow fever," probably typhoid, in 1827, in the thirty-eighth year of his age.

In 1825, when John Shaw was completing his supplementary treatise on distortions of the spine, Jacques Delpech, Professor of Clinical Surgery in the University of Montpellier, had bought $3\frac{1}{2}$ acres of sylvan land, and, giving his fertile imagination a free hand, erected on it an orthopædic institution. The building he designed was a Boccaccio's dream such as the world had not seen before. We had reason to visit Montpellier in a previous chapter to see Jacques, in the year 1816, when he was in the thirty-ninth year of his age and fourth of his professorship. He then introduced to the surgical world a double innovation—the subcutaneous method of operating, and the section of tendons as a means of treating deformities. The reason for our present visit, is to see him apply exercises and gymnastics in a wholesale manner in the treatment of deformities, particularly of spinal curvature. All his life long he kept a sympathetic eye on what was happening among English surgeons. In particular, he commended the "scientific conceptions of the celebrated and clever Dr. John Shaw."

It is worth our pains to note the steps which led Delpech to mortgage his worldly means in realising an orthopædic dream. In 1793—the year of Hunter's death—he became attached to the Army of the South—a mere lad of 16, acting as a surgical dresser. He discovered, as Hunter had done, that the surgeon must base his treatment on a knowledge of function—or, as he phrased it, of "animated anatomy." To him, as to Hunter, a healing wound was a vital field in which every process had to be noted and explained on a physiological basis. If he had not Hunter's depth of insight he had, what Hunter lacked, a glowing enthusiastic imagination and a fluid seductive power of speech. Both had a lust for new knowledge, but while John Hunter was inclined to treasure his stores, Jacques Delpech threw the best he had to the four winds with the open hand of a profligate. Hunter and Delpech had the faculty of making their clinical cases yield physiological secrets. We see that illustrated by the manner in which Delpech became drawn to the study of orthopædic cases. In 1803 Antonio Scarpa, Professor of Anatomy and Surgery in the ancient University of Padua, published the first accurate account of the state of the bones, ligaments, and muscles in cases of club-foot. Delpech studied that work when it appeared, and two cases which he had observed clinically gave him a key to the mechanism which produced the deformity. One was a soldier in whom there was paralysis of the muscles which dorsiflex and evert the foot, owing to a wound of the external popliteal nerve. In the course of time Delpech witnessed the soldier's foot gradually assume the characteristic club-form. Delpech realised then that club-foot must be the result of the unopposed action on the part of the sound muscles of the leg—the muscles which plantar-flex and invert the foot.

In another case, where the muscles of the calf were thrown into a state of spasmodic contraction, due—at least so Delpech believed—to an irritative lesion of the sciatic nerve in the ham, he again saw club-foot produced. In that case he supposed that the deformity was due not to a negative action of the dorsi-flexors, but to a positive action of the plantar-flexors. From these and other cases he drew the important inference that the normal poise of every part of the body is the result of a balanced action between opposed groups of muscles, and if that balance is upset then deformity results. The surgeon's business, he held, was to cure deformities by restoring the muscular balance. Like Hunter, he gathered sound physiological knowledge by the bedside. There is another important point of similarity between the two men. Hunter bought two acres of land near London to set up a menagerie—an experimental garden—in which he might gather knowledge. Delpech bought his 3½ acres with a somewhat similar object in view; into his orthopædic institution he hoped to gather cases of deformity which might be studied, treated, and noted for the benefit of science, as well as for the welfare of patients. When Hunter died his fortune was locked up in his specimens; when Delpech died at the hands of an assassin in 1832, in his fifty-fifth year, his widow had to sell her jewels and trinkets to buy the necessaries of life. Delpech was the pioneer of physiological surgery in France.

Delpech's Orthopædic Institution opened on the great avenue or high road which leads from Montpellier to Toulouse—his birthplace. On entering the gateway, the visitor found himself in the "Grande Cour," the main block with its verandah front facing him; the wing on the left contained the chapel, the winter gymnasium, while in the block on the right there were accommodated the baths, lavatories, and receiving-room. Behind the institution lay a "jardin anglais," a maze of paths amongst shrubs and flower beds with, at its distant end, the great open-air or summer gymnasium, surrounded with trees and furnished with a great "flying bridge." Beyond the English garden lay the orchard with its winding paths in which were placed the swimming school, the cradled see-saws, the "perche horizontale," "plans inclinés," and the "corde horizontale." In a corner of the orchard there was the "salle de la bassécule," of which Delpech gives us a glimpse; we are shown groups of women clad in pantaloons with loose coats bound at the waist with leathern girdles, disporting themselves and strengthening their muscles on a magnified form of see-saw. Here then, and on a great scale, is the first deliberate attempt to apply gymnastics to the treatment of deformities of the human body.

We have seen that Delpech was drawn into the study of deformities through his early interest in club-foot. At the time he designed his institution, and particularly when writing *l'Orthomorphie* in 1827 and 1828, spinal deformities had come to occupy his main interest. He wished to rescue such cases from the hands of charlatans who failed to distinguish curvatures due to tubercular disease from those static deformities which resulted from a weakness of the spinal musculature. It was mainly for the treatment of such spinal cases that his institution was designed. In Delpech we have an ardent advocate of "action" as opposed to "rest" in the treatment of deformities. He held that flat-foot and scoliosis were identical conditions: they were deformities due to the breaking down of normal muscular supports; ligaments and bones were but secondarily concerned. When a case of lateral curvature arrived at the institution he first tested the permanency of the deformity by suspending the patient from the head, but whether the deformity disappeared under suspension or not, he warned the parents or guardians of the patients that time, patience, and perseverance were necessary; treatment would extend over one year at least, perhaps two. On entering, the patient was given a month to become accustomed to a hard mattress and new surroundings. Then commenced a prolonged course of spinal extension, alternated with spells of graduated exercises designed to strengthen the muscles of the back. For fourteen hours out of twenty-four the spine was extended, as the patient lay supine in bed, by counter-traction applied to the head and to the pelvis. Lateral pressure was also applied to the convexities of the curvatures. The extension-bed Delpech regarded as merely an accessory in the treatment; the essential part was the daily period of gymnastic exercises, which he graduated and modified to meet the needs of each individual case. His exercises were chiefly balancing or climbing in character. Delpech gave his imagination a free rein in designing an exceedingly great variety of apparatus, for he was well aware that it was only by ringing a succession of changes that he could maintain the patient's interest in the laborious efforts needed to effect a cure. I cannot resist describing his modification of an apparatus originally designed by John Shaw to prevent young pianists from adopting a faulty posture and developing spinal curvature while at practice. A lady is seen seated at a piano with her back towards us. Her head is encircled by a horizontal band. Behind the piano, and facing the lady, is a column with a swan mounted on the top, holding a pulley in its mouth. Through this pulley passes a cord which connects the band on the lady's head with a weight placed within the column. As long as a proper erect position is maintained, the weight exercises a moderate tension—such as calls the muscles of the back into a degree of action sufficient to counterbalance the tension of the weight. The tension is sufficient to keep the muscles which maintain the spine erect on the alert. They are exercised by having to act against the resistance of the weight.

Unfortunately, the degree of success which attended this great experiment of Delpech we do not know. He died just when it was becoming possible to judge how far his methods were superior to those they were expected to supplant. We do know, however, that in the earlier decades of the nineteenth century and in the South of France one of the ablest surgeons of his time was convinced that gymnastic exercises constituted the best means of remedying certain deformities of the human body. On the other hand, he rejected the use of frictions, rubbings, massage, and oriental baths in the treatment of deformities.

From the South of France we proceed to the South of Sweden to trace the beginning of a movement in the evolution

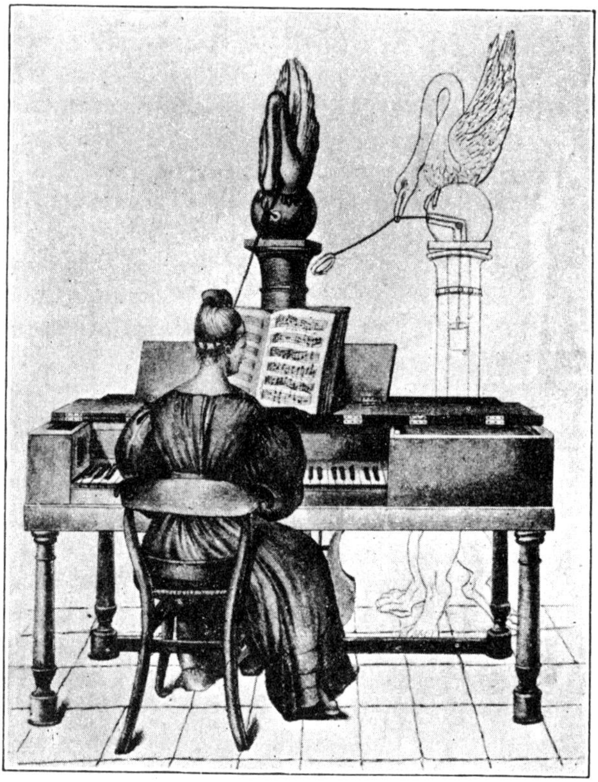

FIG. 1.—Delpech's apparatus for keeping the back straight and head erect. It is a modification of an invention by John Shaw.

of gymnastics, which, in its ultimate effects, was destined to produce a world-wide influence. Pehr Henrik Ling, the founder of the Swedish system of gymnastic exercises, and Jacques Delpech of Montpellier were contemporaries. Ling was born in 1776, son of a scholarly country parson ; Delpech, the son of a master of languages, was born a year later. Delpech was bred to Medicine and became interested in gymnastics later in life, as a means of treating deformities. Ling was bred to Theology, and in his maturer years, through his interest in gymnastics, strayed into the outskirts of the field of Medicine. Ling, we shall find, is a practical physiologist of a new kind—one who gained his knowledge, not in the experimental laboratory or lecture room, but in close observation of the movements which we perform and the postures which we assume in the performance of our daily duties.

In the year 1805 we find Pehr Ling, scholar and poet, professional teacher of fencing and gymnastics to the students of the University of Lund. At the age of 29 he has settled down to earn a livelihood in a definite manner after an eight years' sojourn in the Bohemian quarters of Europe. In teaching the art of fencing—nay, of any of the skilled feats or arts of human life—he came to see that all depended on the mastery of muscles by the pupil's brain; every muscle must respond instantly and exactly to the brain's demand. He found that the pupils who came to him had learned to stand, walk, run, jump, sit, stoop, climb, and all the ordinary acts of life in a semi-unconscious, almost reflex manner. When he came to teach them the art of fencing he found that the special movements which he wished his pupils to carry out were impeded and clogged by being yoked, through use and wont and habit, to other and unnecessary movements. His pupils had already formed their vocabulary of movement and forgotten the alphabet of simple motions out of which complex acts are framed. They had forgotten the alphabet of movement and therefore could no longer coin new words. Ling therefore resolved to teach them the alphabet of the movements of the human body in a systematic manner, and having mastered their letters anew, they could then form new words, new movements, as they desired. He elaborated for the first time an alphabet of movements; that alphabet is his system of "free exercises." He dissected, not the muscles, but the movements of the body; his exercises cover the whole keyboard of our movements—of foot, leg, hip, hand, arm, shoulder, trunk, and head. When the pupil had mastered the alphabet he passed on to easy words—the combination of two or more of the simpler movements, then to more difficult words, and last to sentences—the finished acts, sports, and games of life.

In order that he might understand the human machine he studied muscles, nerves, and the brain in the dissecting room at Lund. But it was as an anatomist of movement, rather than of muscles, that he excelled. He considered that a sound education should include a complete and exact dominance of the muscles by the will. Exercises and gymnastics had not only developed his own muscles, but given him health and strength as well. Having the irresistible impulse of the true missionary he could not rest until he had impressed the value of his discovery upon his countrymen. He saw a new era dawning for them—a period of robust and happy health—if his discoveries and teaching were adopted. His exercises not only developed the body, but they preserved its soundness. Exercises had improved his own health; he had suffered from a rheumatic lesion of some kind which affected his joints, but under his own treatment he had regained health. Presently he elaborated a system of medical gymnastics, including rubbing, kneading, and all the various forms of manipulation by which friction and intermittent pressure can be brought to bear on the living human body.

It was a purely empirical system he introduced; the evidence of its efficacy resting on experience—the results which followed its application. He wrought cures and made converts. Ambroise Paré said, "I dressed the wound; God healed it." But apparently neither Ling nor his disciples took such a humble point of view. It would be asking too much of human nature to expect that the medical practitioners and University professors of Lund should receive the revolutionary proposals and practices of their fencing master with open arms, and hence we find Pehr Ling in Stockholm in 1813 appealing to his friends and to the Government to give him the means of founding a "Central Gymnastic Institution," where he might impart his discoveries to the teachers of his country. His efforts were successful, and when he died in 1839, at the age of 63, he left behind him a permanent organisation and a devoted band of pupils to carry on his work. From that centre in Stockholm his teaching has spread into the playgrounds, barrack yards, and hospital wards of every civilised country.

We have been searching Europe of the nineteenth century to discover the principles and practice of the men who sought to repair the human machine by throwing it into action, sometimes by active, and at other times by passive means. It must be owned we have discovered practice rather than principle. Anyone who examines critically the writings of the modern representatives of the Swedish school [1] will discover that Ling's system of "medical gymnastics" has grown into a bewildering code of arbitrary and empirical practices which can be justified only by the benefits which are said to attend their application. Now there are many methods of treatment which cannot be explained by the physiologist and yet are justified by their curative success. As assistant to a country practitioner I have seen the pain of lumbago, in case after case, instantly relieved by dry-cupping, and yet we cannot explain how the effect is produced. But if we adopt such empirical practices we have to make doubly sure that the results obtained are due to the means we have applied. We must ever keep in mind that all living flesh is endowed with an inherent and marvellous power of repair and of recovery, and if our practice has no known rational basis, then it is our first duty to determine how far the means we employed are accessory to the cure which has taken place. We return to Louis Pasteur's pertinent query—Where is your control series of cases? We search the literature of the Swedish masseur in vain for such an assurance.

We shall turn to the modern writings of Dr. Mennell [2] to discover the best case which can be made out for regarding massage as a scientific practice. In the closing years of last

[1] See *Handbook of Medical and Orthopædic Gymnastics*, by Anders Wide, M.D., 1902. *The Elements of Kellgren's Manual Treatment*, by Edgar F. Cyriax, M.D., 1903.

[2] *Massage: Its Principles and Practice*, by James B. Mennell, M.D., 1917.

century, while a student at St. Thomas's Hospital, London, he worked under a house surgeon who applied the practices of Lucas-Championnière in the treatment of fractures of the fore-arm. In time Dr. Mennell also became an ardent admirer and disciple of the French pioneer of massage, and we can best follow the rationale of his practice from the narrative of one of his earlier cases of fracture of the lower end of the radius: "Finding my patient still seemed to see each attempt at movement and to resist it accordingly, I next tried passing my hand up and down—not with any idea of rubbing the part, but merely to effect a distraction. Great was my surprise to be greeted by, 'Oh, doctor, that is lovely! do go on.' I went on, and I soon discovered that the amount of movement that I had previously attained by the exercise of considerable force and at the expense of much pain to the patient was now procured painlessly and without any attempt at force." By gently rubbing the injured and neighbouring parts, Dr. Mennell had produced (1) a sedative effect—the pain was diminished; (2) the swelling decreased—the lymph and blood circulation of the part had been improved; (3) the muscles became soft and relaxed—they no longer pressed on the fractured ends of the bone and the torn tissues. In seeking for a rational explanation of these effects Dr. Mennell is justified in his contention that they are probably obtained by playing upon the reflex mechanisms, situated in the spinal cord, which control the state of the muscles and the condition of the blood and lymph circulation of the part. Of the reflex mechanism which controls muscles we have a definite knowledge, and we have reason to presume that the circulation of blood and lymph of every part of the body is under the control of special reflexes. We can best account for the swelling and the impeded circulation, which are the invariable accompaniments of injury, by supposing them to be reflex effects excited by the violation of the tissues. Gentle massage movements apparently effect a restoration of the normal control mechanisms. There are also direct effects obtained—the tissue spaces, containing exudates, are compressed; the blood vessels are squeezed and the blood distributed. It is also possible that the metabolism of the tissues—their "power of action"—is also roused. Massage will break down the freshly formed tender adhesions, but under certain circumstances it will also increase the damage already done.

The modern masseuse can appeal to Aristotle and to Nature for support. Aristotle taught that "movement is life." We knead our limbs to refresh our tired muscles; we rub our tired eyes to give them relief; the dog applies massage to his injury by his tongue. But Nature also brings reflexes into action to keep the injured limb from being used and to thus give it rest. An appeal to Nature will support the actionist with one hand and the passivist with the other.

Now, when we listen to the teaching or watch the practice of the followers of Lucas-Championnière we marvel that it was possible for surgeons at any time to have been so blind as to adopt "rest" as the principle in their treatment of injuries. Thomas, of Liverpool, however, had tried both action and rest as means of treatment. He had—what few, if any, of his contemporaries or successors can lay claim to—a series of control cases by which he could carefully measure the effects which attended each form of treatment. His final opinion was that absolute, uninterrupted, and prolonged rest so applied as to support the muscles and leave the circulation free, gave the best results. Nay, he formed the opinion that the swelling, the disturbed circulation, were not hindrances to repair and to be got rid of; they were, he supposed, part of the mechanism which had been evolved for effecting repair. When we see how elaborately and cunningly the living body has been fitted to meet the contingencies of life, it is not too much to suppose that in taking a beneficent view of the swelling which attends injury, Thomas was in the right.

When we come to the application of massage and exercise to assist in the later stages of recovery, we are outside the region of doubt or of dispute. But in the application of massage, exercises, and gymnastics there is one condition which is absolutely essential for success and which is often lost sight of: the patient himself or herself must be an active, not a passive, agent in the cure. The medical officer, often unconsciously, gives the disabled man the impression that he has brought to the bedside the means of cure, and that all the patient has to do is to passively submit and allow a miracle to be performed. Ling never made that mistake, and never taught it; no one knew better than he, that in the recovery of movement, no matter what the nature of the damage to the body may have been, there was only one active agent which could assist in the recovery, and that agent was the patient's will and brain. There is no miracle in the art of healing except that which is wrought by the power of repair inherent in living flesh; all the rest is hard work—hard thinking on the part of the medical man and constant and intelligent effort and co-operation on the part of the patient. "Any method of treatment," writes Colonel Deane,[1] "which lacks the essential factor of stimulating and encouraging the man's own power of will-to-do, which has been inevitably weakened by the stress through which he has passed, stands self-condemned. This factor of volition is vital. The man must do this himself, and exercise his own volition and put his own motor centres in action. A fraction of movement obtained in this way is infinitely more valuable than a greater amount gained by passive methods." Thus we see that psychology also has a place in the armamentarium of the orthopædic surgeon.

[1] *Gymnastic Treatment for Joint and Muscle Treatment*, by Colonel H. E. Deane, R.A.M.C., Oxford Medical Press, 1918.

HENRI-LOUIS DUHAMEL DU MONCEAU,
BORN 1700; DIED 1782.

CHAPTER XIV

THE FOUNDATION OF OUR KNOWLEDGE OF BONE GROWTH BY DUHAMEL AND HUNTER

In this chapter are reviewed the experiments and observations which gave us the foundation of our knowledge of the manner in which bones grow. John B. Belchier, the surgeon who discovered that madder, when eaten by an animal, stains its bones red and leaves all the other tissues of the body uncoloured, was on the staff of Guy's Hospital. He was born in 1706, and died in 1785. Duhamel, who was a French squire, discovered that madder only coloured bone matter as it is being laid down, and therefore provided a certain means of telling how bones grow. He came to the conclusion that periosteum was the mother-tissue of bone. He was born in 1700, and died in 1782. When Duhamel was carrying out his experiments Albrecht von Haller, a Swiss born at Berne in 1708, and dying there in 1777, was Professor of Anatomy in the University of Göttingen, and from his experiments and observations came to the conclusion that bone was not formed by periosteum, but was a product of arteries which could lay down bone anywhere—within cartilage or beneath the periosteum. John Hunter (1728–1793) adopted the madder method and many others, for investigating the life and growth of bone, and discovered that in growing bones there are always two processes in operation—one of deposition of new material, and the other an absorption of the old, by the co-operation of which every bone is being constantly remodelled as it grows.

SOMETIME in the summer of 1736, a calico printer of the City of London entertained to supper John Belchier, a promising young and inquiring surgeon, and, all unwittingly, introduced him at the same time to a method of unravelling some of Nature's greatest secrets. The method to which Belchier was thus introduced was that of vital staining, one by which certain elements of the body can be picked out by distinctive colours while they are still living. The calico printer was an economical man, and used the madder-soaked bran from his dye-vats to feed pigs. The pigs he used to feed his friends. It was a joint of this madder-fed pork which was placed before the young surgeon, then in his thirtieth year and just elected to the staff of Guy's Hospital. It was but natural that a young man, already a fellow of the Royal Society and a contributor to its *Philosophical Transactions*, should have his curiosity aroused by the ruddy colour of the bones of the madder-fed pork, and that he should resolve, on returning to his home, to make further inquiries before bringing this strange matter to the notice of the Fellows of the Society. He desired, first of all, to make certain that it was madder and no other substance which stained living bones, and hence he began to feed some of his fowls with madder. One cock, subjected to this treatment—fed by force, for fowls refuse food mixed with madder—died at the end of a sixteen days' course; the bones were stained, even those of the densest structure. Early in the autumn he made a communication to the Royal Society, which was subsequently printed in its *Transactions*.[1]

Having duly recorded this strange action of madder on bones, Belchier felt that his inquiry was finished, and left the matter there. The man who was to show that Belchier had

[1] *Phil. Trans.*, vol. xxxix. p. 287.

discovered a means of unravelling the complex manner in which bones grow was a remarkable French squire—Henri-Louis Duhamel, Seigneur du Monceau—but of him more anon. For Belchier there was no problem of bone growth. His master and teacher, the great Cheselden, knew all there was to be known about bones—particularly so far as concerned those of the human skeleton. Bones "are covered by a fine membrane, which upon the skull is called pericranium, elsewhere periosteum. It serves for the muscles to slide easily upon; it is everywhere full of small blood vessels, which enter the bones for their nourishment." Bones "grow by the continual addition of this ossifying matter; they increase till their hardness resists a further extension; and their hardness always increasing while they are growing, the increase of their growth becomes slower and slower until they cease to grow at all." "In a fractured bone, in which the same kind of matter which ossified the bones at first, is thrown out from the broken ends, there is formed a mass of callous matter."[1] It is extremely important that we should note these points: that periosteum was a vascular membrane which supplied nourishment; that bones grew like other tissues—by a natural filling up and expansion of every interstice; and that, when broken, the fractured ends threw out a mass of ossific callus. That was the state of knowledge as regards bones when Duhamel and Hunter commenced their investigations.

Our story now shifts from London to France—to the well-managed estate of Monceau, where Duhamel, a confirmed bachelor, and his maiden sister spent busy and studious days. Duhamel belongs to that rare type of man which is represented by Stephen Hales and Benjamin Franklin—men born with a natural aptitude to play the part of detectives amongst Nature's secrets. They wrung them from her by experiment. For Duhamel there was but one sure source of knowledge—observation and experiment. He had spent his youthful days in Paris as a student—a student of law; but he had attended Winslow's lectures, and at an early date had begun

[1] *The Anatomy of the Human Body*, by William Cheselden, 6th edition, 1741.

investigations relating to agriculture which brought him into touch with the Academy of Science and the foremost researchers of his time. At the age of 27 he was elected a member of the Academy. We make his acquaintance in 1739, when he was 39 years of age, and had just made his first communication to the Academy, "Sur une Racine qui Teint les Os en Rouge." He was then living on his estate. A year or two before, Hans Sloane had drawn his attention to Belchier's paper. His curiosity was aroused, and, following his usual custom, he submitted Belchier's observations to the test of experiment. He fed fowls, turkeys, pigeons, and pigs with madder, and found that their bones, and only their bones, were stained red; he noted that only certain parts of the bones were stained, and that those of young animals coloured more readily and deeply than those of old animals.

In his communication of 1739 he had really added very little to Belchier's observations. It was merely a preliminary note. His next communication appeared in 1741, and we see he had been hard at work. He had become interested in bones; he knew well how a broken twig became mended, and he wondered how repair was carried out when a bone was broken. He made experiments, and found that the periosteum in the neighbourhood of a fracture became greatly swollen, and served as the chief agent in producing the repairing callus. The lining medullary membrane, or internal periosteum, assisted. By feeding the subjects of his fracture-experiments with madder, he could trace the formation of the newly-formed bone, and found that its chief source was the periosteum. He saw nothing to support the prevalent belief that the callus was produced by a "juice" secreted by the broken ends of the fractured bone. He had now realised that the bony matter which was stained red was that formed during the period of madder-feeding. Only the newly deposited bone was stained, and therefore the staining gave him a clue to the manner in which bones grow.

His communications of 1742 and 1743 show us that his further experiments had been attended by a great success.

He had discovered an early mistake. If, after giving a course of madder food, and then a course of ordinary food, a pig was killed, he observed in his earlier experiments that nearly all redness had gone from the bones. He was thus led to believe that the stain was only temporary. But on laying such bones open, he observed his mistake—the madder-coloured layer was only covered and hidden by an unstained surface layer, laid down after the madder diet had been suspended. He found, by alternating a madder diet with an ordinary one, that he could obtain alternate red and white rings or layers on the circumference of the bone, the latest-formed layer or plate being on the surface of the bone immediately under the periosteum. Bone therefore grew as wood did—by the superimposition of layer on layer or plate on plate; and the source of the new layers was the periosteum. The deepest stratum of the periosteum apparently served the same purpose as cambium did for growing wood. In this way Duhamel came to regard periosteum as the maternal tissue of bone; the osteogenic function of periosteum is Duhamel's discovery.

He was faced by a great puzzle. Bone grew in thickness only by superficial additions laid down lamina upon lamina. He noted that the medullary cavity also kept enlarging as the bone grew; in the pig's tibia the cavity kept increasing until the animal was six months old. He could conceive only one way in which the medullary cavity could undergo an enlargement—namely, that as the superficial layers were laid down, there was also an expansion of the shaft-cylinder as a whole, an expansion which enlarged the medullary cavity. He sought to prove the truth of his theory by a new kind of experiment—one he had practised on growing branches. He encircled the shafts of growing bones with rings of silver wire, and in the course of time found, as he expected, that they had cut their way into the medullary cavity. He believed that such a result could be obtained only by an expansion of the shaft. We shall see presently how John Hunter explained the result of such experiments.

Duhamel was familiar with the fact that bench marks cut

on the stems of trees maintained a constant level; growth in the trunk did not alter the level of the step or bench mark which had been cut many years before. He applied a similar method to bone; he cut "bench marks" on the shafts of growing bones, by drilling holes at regular and measured distances, inserting within them silver stylets to keep them open. After a definite period he killed the animal, measured the distance between the bench marks, and found they were unaltered; there had been growth in the total length of the bone, but the distance between his marks on the shaft remained the same. Therefore he concluded that all growth in length must take place at the extremities of bones. At what particular part of the extremity, he makes no mention; he observed the epiphyseal lines, but left them unstudied.

Duhamel has been unfortunate in his commentators; with the exception of Flourens, who repeated and extended his experiments in 1842, all have noted his mistakes, and forgotten that he was the man—one not trained to Medicine—who discovered that there was a problem of bone growth, and showed the world how such a problem should be solved. He turned the chance observation of an English surgeon into a powerful instrument of research. He showed that bone is stained with madder only as it is being laid down. He showed that a bone grew in thickness by circumferential deposition of plate upon plate; that a long bone grew in length at its extremities; that the deepest layer of the periosteum was the maternal tissue of bone.

Our story now shifts back to London. Duhamel's researches were carried out while John Hunter was spending an untrammelled boyhood on his father's farm at Long Calderwood. But the storm which Duhamel's discoveries had raised was still in full blast when in 1748 Hunter joined his brother's school in Covent Garden at the age of 20. The great Haller and his pupils had directed a flood of destructive criticism on Duhamel; Haller's name stood high in William Hunter's school; everything he wrote and did received the most respectful attention. If anyone was John Hunter's mentor more than another it was Haller. "How could any anatomist think," said Haller, "that bone is formed only by periosteum?" Why, one had only to look at the lower end of the thigh-bone of a child at birth to see the theory was untrue; there one could see arteries perforate the epiphyseal cartilage, and at their growing ends observe a point of ossification start into being—far removed from the periosteum. In Haller's opinion arteries were the depositors and builders of bone; he believed they could form bone anywhere within the limits of the periosteum. One has only to look at Hunter's preparations, or glance at scattered passages in his writings, to see that it was Haller, not Duhamel, who influenced him. He prepared a series of specimens to show ossific centres forming in the patella, sternum, and metacarpal bones of the calf, and in young epiphyses of human bones. These preparations show us that for him ossification was a function of arteries; ossification, he observed, always broke out at the terminal meshwork of an ingrowing leash of blood vessels. Haller had repeated Duhamel's madder-feeding experiments, and again studied the repair of fractures, and concluded that the periosteum took no essential part in the formation of the repairing callus—the callus was formed by the broken bone; nor did the periosteum take any part in the formation and growth of bone. In Haller's opinion the periosteum was, as all surgeons then believed, merely a vascular covering to serve for the nourishment of bone. Thus, with the appearance of Haller and Duhamel, we have two schools brought into existence—one which regards periosteum as osteogenetic, the true source of bone, and another which regards it as devoid of any bone-forming power whatsoever. After a century and a half of inquiry and discussion we still have these two schools.

To which of the two Hunter belonged there is no shadow of doubt. For him the normal periosteum had no bone-producing power; it was the passive vascular membrane which surrounded and nourished bone. Yet he had seen the periosteum take on an "ossific disposition" time after time under the most varied conditions of disease and experiment. He believed with Haller—and we shall see the pains he took to verify his belief—that arteries, and only arteries, could form bone, and if periosteum should, under abnormal circumstances, become a site of bone-formation, then that, to his mind, was merely evidence that its arteries could assume a bone-building power.

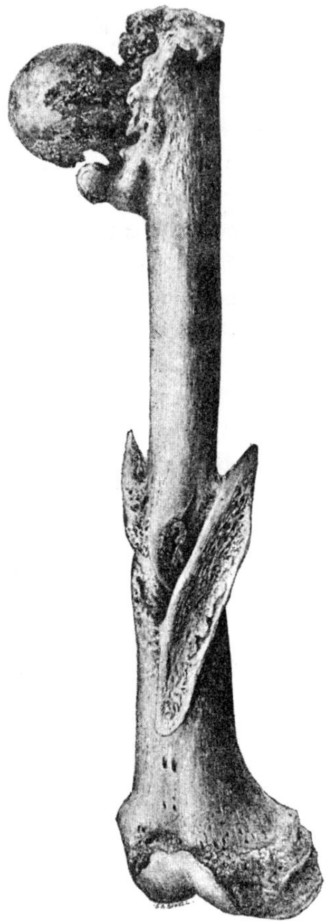

FIG. 2.—Drawing of the fractured femur which Hunter was in the habit of showing to his class to demonstrate that a large fragment of bone may be detached from its periosteum and yet live and again form a union with the shaft. (Drawn by S. A. Sewell.)

Hunter was well aware that a fragment of living bone could be detached completely from its periosteum and yet retain its "vital principle" and survive (see Fig. 2). "In fractures," he said, "adhesion of the detached splinters also takes place . . . and this takes place not only in those which are attached to the soft parts, but even in such as are entirely loose. Therefore these pieces must retain the living principle, and probably only this, while those that remain attached have probably more. I have never examined a compound fracture without finding some of these loose pieces, which shows they must be common. Their union must be similar to that in the transplanting of teeth."[1] It is clear from the above statement that Hunter knew that isolated fragments of bone could be made to serve as bone grafts. The specimen which he was in the habit of showing at his lectures, to clinch the argument that a large fragment of bone might be detached from its periosteum and yet live, is illustrated in Fig. 2.

[1] *Collected Works*, vol. i. p. 502.

It is a fractured femur, showing a splinter 5 inches in length which has been so completely detached that its position is reversed; yet it now forms a solid part of the shaft at the point of union.

We come now to trace the steps which led Hunter to recognise and explore one of the most remarkable of all the powers possessed by living tissues—the power to absorb and remove parts of their substance. The first step occurred in 1754, when he was in his twenty-sixth year and assisting in the management of his brother's anatomical school. He had then commenced his researches on the jaws and teeth, and was struck by the fact that, when a tooth was extracted, its socket disappeared. Thousands before him must have noted the fact, but they passed it by as self-explanatory. When he came to examine the manner in which milk teeth were shed and permanent teeth came into place, he noted that the roots of the teeth about to be cast out were eroded, and that the sockets in which they were contained were disappearing. He noted also that the last milk molar lay at the root of the ascending ramus of the mandible. But, in the adult jaw, room had been found for the additional three permanent molars between the site of the last milk molar and the ascending ramus. How could such additional dental space be obtained unless the mandible had been remodelled during growth by a dual process of addition and abstraction? He was faced with the same problem when he came to explain how the femur of a newly born child is transformed into that of the adult. Mere deposition of new bone at the upper epiphyseal line only adds to the upper surface of the neck of the bone (see Fig. 4). As the femur grows, its neck ought to increase rapidly in depth and breadth, unless deposition on its upper surface is accompanied by a simultaneous process of absorption along its lower surface. It thus became clear to Hunter that, from the infantile to the adult stage in the growth of a thigh-bone, the femoral neck was being continuously remodelled. Growth in bone, he concluded, entailed two distinct processes—one of deposition and one of absorption. That was Hunter's discovery. There was no need to suppose, as Duhamel had done, that the shaft of a bone actually expanded as it grew; the medullary cavity enlarged because, as new bone was deposited on the surface of the shaft, old bone was removed from its interior by absorption.

Soon after Hunter commenced his experimental farm at Earl's Court in 1764, he resorted to Duhamel's method to test the proof of this theory of bone growth. He fed two young pigs on a madder diet for a month; he killed one at the end

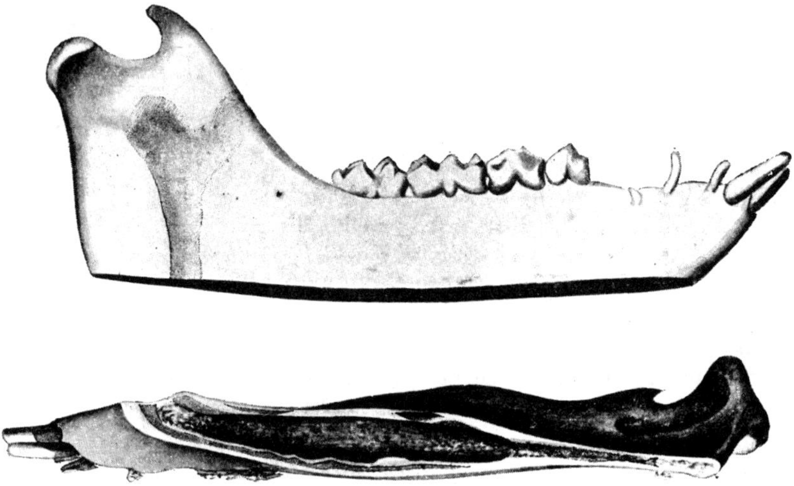

FIG. 3.—Copy of Hunter's drawing to show the manner in which growth takes place in the mandible. The drawings represent two views of the right half of the mandible of a pig, which was fed on a madder diet for a month and then for a month on normal diet. The outline of the ramus at the end of the madder period is shown by the colour. (Copied by S. A. Sewell.)

of the month; the other he kept for an additional month on ordinary food before killing it. Twenty preparations from these two animals are still preserved in the R.C.S. Museum, but the colouring has almost gone out of them. Fortunately, there are preserved accurate coloured life-size drawings of the jaw and femur of the second pig—killed a month after the madder diet had ceased (Fig. 3). Hunter found the appearances presented by these two bones to answer his expectations in the most exact manner. What had been the condyle and posterior border of the mandible during the madder period were now included in the substance of the ramus—buried by the new condyle and new posterior border which had been added in the non-madder period. From the anterior border of the ramus the madder-stained bone was almost completely removed by absorption. He found that the thigh-bone (Fig. 4) told the same story: what had been the upper surface of the femoral neck during the madder period was now surmounted by new bone, but on the lower aspect of the neck there was no new deposit, but a clear indication of absorption. As the result of his madder-feeding experiments, he concluded that living bone "is constantly changing its matter," and that absorption is as essential a part of bone growth as deposition.

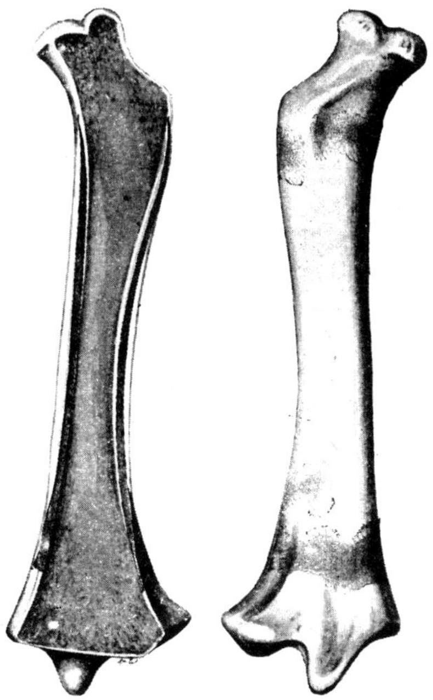

FIG. 4.—Two views of the right femur of the same animal as that from which Fig. 3 was taken. The position of the epiphyseal ends at the termination of the madder period is shown.

The recognition that absorption is an essential factor in the process of bone growth brought Hunter face to face with another problem during those youthful years he spent in his brother's dissecting room. Bone was laid down by arteries, but by which system of structures was it absorbed and carried into the circulation? At the period he was making his first investigations on the teeth—from 1754 to 1758—his brother William was proclaiming as a new discovery the absorbent function of lymphatic vessels: lymph vessels were, he said, of the same order as lacteals; therefore they had the power of absorption. In 1758 we find John Hunter instituting experiments to ascertain if veins had the power of absorption, and apparently was satisfied when he thought he had proved that they had no such power. There remained only one system of structures which could remove substances from the tissues — the absorbents. "If any solid part of the body undergoes diminution, brought on in consequence of disease, it is the absorbent system that has done it; they are the thieves!"

After having settled to his satisfaction that lymphatic vessels were the agents by which bone was absorbed and modelled, a much more difficult problem presented itself—one which shows us how deeply Hunter sought to pry into Nature's secrets. He saw that the living bone of empty tooth-sockets was removed. He was certain that lymphatic vessels were the agents by which its removal was effected. But what circumstance set the lymphatic vessels to work on the walls of the empty tooth-socket? Why should the living bone of that socket submit? In answering these questions, Hunter seems to offer us a stone rather than bread. "The remote cause of absorption of whole and living parts implies the existence of two conditions, the first of which is a consciousness, in the part to be absorbed, of the unfitness or impossibility of remaining under such circumstances, whatever they may be, and therefore they become ready for removal, and submit to it with ease. The second is a consciousness of the absorbents of such a state in the parts. Both these concurring, they have nothing to do but to fall to work."[1]

To anyone whose imagination has not spanned the wide gulf which separates the simple hydra—Hunter's "polypus"—from the highly organised tissues of the human body, Hunter's solution of the problem of absorption will sound like one culled from the land of dreams or of metaphysics. But to one who has spanned that gulf, as Hunter had done, and recognises that every living particle of the human body retains something of that quality of the hydra which we name consciousness, it will be realised that Hunter's answer is a real and true one. In another chapter[2] we shall have occasion to

[1] *Collected Works*, vol. i. p. 255. [2] Chapter XVIII.

examine a law which is applied to explain the manner in which bones shape themselves—Wolff's law—and we shall find that, unless the bone-forming and bone-absorbing cells possess just such a property as Hunter ascribed to arteries and absorbents, then such a law is without foundation; yet that law is well founded.

No one has given a more vivid or a more accurate description of the manner in which a living bone casts off a dead piece than Hunter. Indeed, it was his study of necrosis and exfoliation that assisted him to realise how important an operation absorption was in the life of bones. I allude to this part of his investigations because we again find him ascribing a consciousness, not this time to the living bone, but to Nature. "Thus, while Nature is busied in getting rid of that part of the bone which is dead, she is laying on additional bone on the outside, the intention of which seems to be that of keeping up the strength of the bone, which would, without this addition, be lessened by the loss of substance. This opinion is, I think, supported by the circumstance seldom occurring in this manner in any bones but those of the lower extremity, which support the animal."[1] We have here formulated for us one of the chief observations on which Wolff's law is based.

We need not review Hunter's investigations of the processes involved in the healing of fractures, further than to ascertain how nearly his observations agree with those of Duhamel. Hunter regarded the vascularisation of the blood-clot, which is effused between the broken ends, as the first important stage in the formation of the repairing callus. He saw that the vessels of the callus may be derived from the periosteum, from surrounding torn membranous or muscular tissue, or from the broken ends of the bone. He believed in the Hallerian doctrine that any arteriole could deposit bone, and it therefore mattered little whether the arteries of a callus came from those of the bone, periosteum, or neighbouring muscle. The deposition of bone, he observed, usually commenced at the broken ends, but separate deposits might also

[1] *Collected Works*, p. 574.

occur in the callus. He thus differed altogether from Duhamel, who regarded the uniting callus as a product or secretion of the periosteum.

Hunter repeated the experiments on which Duhamel based his belief that a bone could grow in length only at its extremities. He inserted two pellets in the tibia of a young pig; when the tibia was fully grown the pellets remained exactly the same distance apart. In the course of growth, if the theory of intercalated growth were true, then the pellets ought to have been pushed apart. The experiment supported Duhamel's contention. Fig. 5 shows the shank bones—tarso-metatarsus—of fowls on which Hunter performed similar experiments. One of the specimens is laid open, and reveals two leaden pellets in the medullary cavity—the distance apart being exactly the same as when implanted at an early stage in the growth of the bird. The other Hunterian specimen, also depicted in Fig. 5, shows two canals leading into the medullary cavity. Bristles inserted into these canals demonstrate that both are inclined towards the upper or proximal end of the bone where is situated the main epiphyseal line of growth. Now, when Hunter made the holes, the bone was 71 mm. long and the distance between the holes 42 mm.; when the fowl was killed, the bone was 91 mm. long and the distance between the holes 48 mm. At first sight it looks as if Hunter had proved that there had been an interstitial elongation of the shaft, for the two holes have separated 6 mm. in the course of growth. Hunter has left no interpretation; but I suspect he had fathomed the matter, and realised that he had discovered the cause of the obliquity of the nutrient canals of long bones. The periosteum is firmly fixed to the epiphyseal line at the upper end of the shaft where all growth in length is taking place. That growth at the proximal end of the bone causes a drag on the periosteum—a drag which decreases in degree as the distal end of the bone is approached. Hence the cauterised holes in the periosteum are dragged, during the growth of the shaft, an unequal degree towards the upper or growing end of the bone, and thus become separated to an increased extent. Sixty years after Hunter's death, Sir George Humphry explained the oblique course of nutrient canals of long bones in the exact manner shown by Hunter's experiment.

The R.C.S. Museum contains a series of specimens which represent one of Hunter's most profitable and illuminating studies of bone growth. Antlers of the fallow deer at all stages of growth—with their summer velvet covering, in their

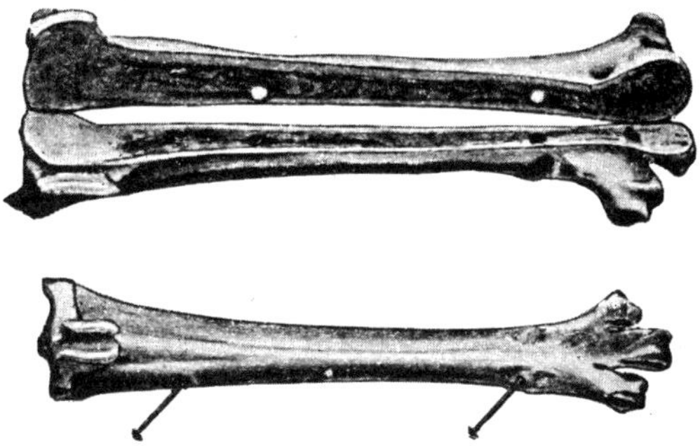

FIG. 5.—Two drawings of Hunterian specimens—both of them tarso-metatarsal bones of fowls on which Hunter experimented.

The upper bone is laid open. The pellets which were inserted in canals of the shaft at an early stage of growth are now in the medulla, but the same distance apart as when inserted in the young bird.

In the lower specimen, two pins are inserted in two canals, to show that their direction is oblique. Hunter made two fine holes by a cautery in the shaft during the chick stage. He found the holes had become oblique canals, and that their external mouths had separated six millimetres in the course of growth. (Drawing by Wm. Finerty.)

late autumn condition, and in the state reached in the early spring before they are shed—have been prepared by Hunter so as to show the rich carpet of blood vessels which marks the site of bone deposition under the "velvet" and at the tips of the tines. Sections at the root of the antler expose the condition of the burr and the state of the bone in all the stages which precede the death and shedding of the antler. He realised that Nature has never given more vivid demonstrations of bone growth, never provided such wealth of opportunities for discovering the secrets of necrosis, as in the lavish annual expenditure she makes in providing deer with antlers. No one looking at the rich carpet of arteries which clothes the surface of the rapidly growing bone would fail to conclude, as Hunter did, that arteries are the principal agents in bone formation. He saw how complex the whole process was—how it was influenced by the state of the sexual glands, and how the arteries of the neck, by some secret mechanism, became enlarged to serve the needs of the growing antlers. We shall see how far the introduction of the microscope changed our point of view sixty years after Hunter's death. Meantime we note that those two men of the eighteenth century—Duhamel and Hunter—laid the basis of our knowledge as regards the physiology of bone; and of the two, Hunter laid the firmer and deeper part of that foundation.

JOHN GOODSIR,
Professor of Anatomy in the University of Edinburgh 1846–1867,
BORN at Anstruther, Fife, 1814; DIED at Edinburgh 1867, aged 53.

CHAPTER XV

RESEARCHES MADE BY SYME AND BY GOODSIR REGARDING THE GROWTH AND REPAIR OF BONES

An exact knowledge of the manner in which bones grow and are reproduced became of great practical importance when surgeons began to excise diseased and disorganised bones so as to save limbs from amputation. Resection of bones as a means of obviating amputation was introduced by Charles White of Manchester in 1768; Henry Park of Liverpool, with a similar purpose in view, initiated the operation of excision of joints in 1783. James Syme, Professor of Clinical Surgery in the University of Edinburgh, defended the Duhamelite theory that bones which had been excised were reproduced by periosteum, while his junior colleague, John Goodsir, Professor of Anatomy, demonstrated that such a theory was untenable, bone being reproduced only by osteoblasts, the essential constituents of living bone.

CHARLES WHITE was the son of a physician in Manchester, where he was born in 1728, and where he died in 1813.

HENRY P. PARK, surgeon to the Liverpool Infirmary, was born in that city in 1744, being the son of a Liverpool surgeon. He studied in London under Percival Pott, John Hunter's senior and successful rival. Park died in 1831.

JAMES SYME, although born in Edinburgh (1799), came of a Fifeshire family. He became Professor of Clinical Surgery in 1833 at the age of 34, and died in 1870 at the age of 71. He was succeeded in the chair of Clinical Surgery by his son-in-law, Joseph Lister.

JOHN GOODSIR was the son of a medical man practising in Anstruther, Fife, where Goodsir was born in 1814. He succeeded the third Monro in the chair of Anatomy in the University of Edinburgh in 1846; he became the martyr of a long ataxic illness and died in 1867, when he was succeeded by the late Sir Wm. Turner.

MY last chapter closed with the death of Hunter in London in the autumn of 1793. We shall now proceed to Edinburgh, to see what two men who made Edinburgh famous in the middle decades of last century—Syme the surgeon and Goodsir the anatomist—can teach us regarding the manner in which bones grow and heal. It suits our purpose, however, before proceeding to Edinburgh, to discover how bone surgery stood in the North of England during Hunter's time.

No one who values a knowledge of the history of surgery and of learning in England can afford to be ignorant of Charles White of Manchester. He was born in the same year as Hunter, 1728; he was a class-fellow of Hunter in the Covent Garden School; through his long life he never wavered in his loyalty to his friend's splendid abilities. He was elected a Fellow of the Royal Society long before Hunter. Cullingworth has assigned to him a foremost place amongst British gynæcologists; Cunningham has given him equal rank amongst the pioneers of British anthropology. He took a leading part in founding the hospitals, colleges, and learned societies of his native city, Manchester.

We propose to make his acquaintance in April 1768, when Edmond Pollet, a country lad of fourteen, presented himself at Manchester Infirmary with an abscess in the left shoulder, the disease having commenced suddenly a fortnight previously. Pus was issuing by several sinuses situated in the pectoral and deltoid regions. If Charles White, then in the prime of his professional career, had followed the standard treatment of the period, he would have amputated the arm; but he resolved to try and save the limb by adopting a novel and conservative method of treatment. Making a deep vertical incision along the deltoid region, he forced the upper end of the dead shaft of the humerus from the wound and cut it off. Later, a further piece of the shaft was removed—altogether four inches of it. The arm was placed in a sling, so that "the weight of the arm might counterbalance the contraction of the shoulder muscles." Movement was permitted, to secure a sufficient degree of mobility at the shoulder-joint. In four months, to Charles White's surprise, the upper end of the humerus had been reproduced; there was only one inch of shortening, and the arm could be freely used. In reporting this case to the Royal Society in 1769, White drew two conclusions: (1) That when a bone was removed, the arteries of the part had the power to renew it; (2) that the operation of resection had manifest advantages over amputation, and deserved serious consideration as an alternative measure.

Charles White's novel operation of 1768 merely involved the resection of dead bone. In 1783, Mr. Henry Park, surgeon to the Liverpool Infirmary, was in a position to announce a further stage in the progress of surgery—the possibility of excising diseased joints as an alternative to the amputation of limbs. We merely note here that each of these innovations in the treatment of diseased bones and joints demanded a more intimate and exact knowledge of the life and growth of bone.

Having thus seen the rise of "conservative surgery" in the North of England in the later half of the eighteenth century, we pass on to Edinburgh to watch the practice and principles of James Syme. In his day Syme held the centre of the surgical stage; but it is only the early part of his career which interests us now—the struggling decade between 1823 and 1833, from his twenty-fourth to thirty-fifth year, when he was forcing his way up the Edinburgh surgical ladder. In 1825 there was sent to him a woman, Christian Laing, age 38, who had suffered from caries (tuberculosis) of the shoulder-joint for six years. Syme was brought face to face with the same problem as had been presented to Charles White fifty-seven years before—should he amputate, or should he resect the joint and thus try to save the limb? He chose the latter, and the woman did well. Resection as a defined operation for the relief of a diseased joint was, as we have seen, first formulated in 1783 by Henry Park, surgeon to the Liverpool Infirmary,

JAMES SYME,
Professor of Clinical Surgery in the University of Edinburgh 1833–1869.

in a letter to Percival Pott; but the suggestion which moved Syme to adopt the operation came from a French source. Almost at the same time as Park was elaborating and applying the operation of resection to joints in Liverpool, Moreau, a surgeon in the North-East of France, was adopting a similar method of treatment, though it was many years later (1816) before his French colleagues recognised the value of his work. The first effort of Syme's youthful career was to introduce this operation into Scotland. In 1831 he published a *Treatise on the Excision of Diseased Joints.* By that date he had done fourteen resections of the elbow-joint. The introduction of a new surgical method invariably makes new demands on our knowledge of anatomy and physiology. The adoption of resection of joints as a routine measure made an exact knowledge of bone reproduction more urgent than ever.

If we would use words in a strictly conventional sense, we must describe Syme as an eminently practical as well as a daring surgeon. To the onlooker he seemed to apply a remarkably sound, common-sense judgment to the various surgical problems which came before him. In reality he was, as all purely practical men must be, an experimental surgeon, trying to discover the best means of treatment by roughly balancing the results which attended one method against those which followed from another. But for one brief period in his life Syme became a rational surgeon. In 1835—two years after he succeeded to the chair of Clinical Surgery in Edinburgh University—he had to treat the case of a young girl who suffered from acute osteomyelitis of the tibia. He had to amputate. When he dissected the leg, he found the shaft of the bone dead, and enclosed by a new stratum of bone which was adherent to, and apparently formed by, the periosteum. To settle for all time that periosteum could and did reproduce bone, he carried out the following well-devised—and, I have no doubt, neatly executed—experiment on young dogs.

In his first experiment he removed 1¾ in. from the radius of both right and left legs; on the right he also removed the periosteum, whereas on the left this membrane was preserved in place. When the dog was killed at the end of six weeks, the missing part of the left radius was reproduced; whereas, on the right, where the periosteum had not been preserved, a gap still remained. He repeated the experiment on a second dog, and got the same result: when the periosteum was removed, a gap remained.

He changed the conditions of the experiment. Taking another dog, he raised the periosteum from a segment of the radial shaft, and separated the periosteum from the bone by introducing between them a ferrule of tin-foil. The periosteum formed a layer of bone on the tin-foil.

Finally, he cut away the periosteum from a segment of the radius, and wrapped the denuded surface with tin-foil. No layer of bone was deposited on the tin-foil, because there was no periosteum covering it.

We have here the Haller-Duhamel controversy again raised, and it is clear that Syme became a convinced Duhamelite.

16

We are now to make the acquaintance of a man who in his intimate and deep grasp of the problems presented by living matter came nearer to Hunter than any other Briton of the nineteenth century. In physical appearance John Hunter and John Goodsir were as different as men can be. Hunter was from the West, of medium stature, alert, and fiery; Goodsir was from the East—from Anstruther in the county of Fife—tall, gaunt, serious, and Calvinistic, with a long, sombre countenance. But Hunter and Goodsir were alike in this: they both had an insatiable capacity for work, and an almost "uncanny" gift of rightly interpreting, from the slenderest of clues, the more secret workings of living matter.

Here, however, we have to do only with Goodsir's investigations concerning the growth and reproduction of bone. Goodsir was fully aware of Syme's researches on the reproduction of bone by periosteum; he had "dressed" for Syme; he was a student in Edinburgh, and apprenticed to Nasmyth the dentist when Syme was rising to the height of his fame. There was every inducement for him to support Syme's conclusions; but the result of his investigations convinced him that the periosteum had no power to form bone, and that there was only one element in the body which could form and reproduce bone—namely, the corpuscles, which he had detected and studied with his microscope as normal constituents in every variety of ossification. Goodsir was a Hallerite of a new kind.

No one can serve so well as Goodsir to bring home to us the transformation which took place in our knowledge of living bone in the decade which followed 1838—when men first came to realise that living matter is made up of units or atoms to which we give the name of cells or corpuscles. We are pursuing a straight path when we lay bare the circumstances which made Goodsir one of the pioneers of the cell doctrine. There are, in the first place, the circumstances of his training. He came of a line of medical practitioners. At the age of 16—in 1830—he was apprenticed to Nasmyth the dentist. He had the most delicate control of his long fingers, and very soon began to apply his manipulative ability to unravelling the delicate tissues of tooth buds and his quick eye to noting the several stages which the various teeth passed from the time their rudiment could be detected in the embryonic jaw until the date of their complete eruption. Such an investigation entailed a collateral study of bone growth. In Knox's dissecting room he was thrown amongst the men who were to be leaders in Science and in Medicine—Edward Forbes, Fergusson, John Reid, and Sharpey. He dressed for Syme and dissected for Alison. In 1835 he became a licentiate of the College of Surgeons of Edinburgh, and returned to Anstruther to assist his father. In three years he was back in Edinburgh in search of a foothold on the ladder which led to the more eminent posts in Medicine. He had to wait until 1841, when he was 27 years of age, before one came—the curatorship of the Museum of the College of Surgeons, at a salary of £150 per annum, out of which he had to pay an assistant. He only held the office for three years; but never did man crowd years with discoveries as Goodsir did these three. They were discoveries, we shall see, that have a very direct bearing on the problems we are exploring.

Goodsir brought to his office not only rare powers of brain and hand, but two qualifications which are absolutely essential for the man who would open a new field of knowledge—a new instrument, and a fresh inspiration. The new instrument was the compound achromatic microscope, which had then been perfected by Lord Lister's father, and such an instrument was Goodsir's companion from his boyhood upwards. The inspiration he drew from years spent in the study of the marine forms of invertebrate life which he dredged from the waters of the Firth of Forth. Molluscs, echinoderms, and sponges presented him with the same problems of life as did the tissues of the human body, but in a simpler form. He had thus acquired the means of spanning the gap which seems to separate the simple tissues of the lowly hydra from the complex textures which compose the human body with much greater accuracy than was possible in Hunter's time. When he came to unravel the obscure and complex process of bone-building in the human skeleton, he found a clue during his study of sponges; they also build skeletons.

Goodsir was a follower of Hunter. In his youth he accepted Hunter's teaching that bone was deposited by arteries and was absorbed and remodelled by lymphatics. The application of the microscope soon shattered these beliefs. His microscopical investigations of the growing epiphyseal lines of bones revealed cartilage cells grouping themselves into serried ranks; he saw, or thought that he saw, these cartilage cells settle in the meshes of the cartilaginous trabeculæ and become "bone-forming corpuscles"; he traced the transformation of the primary spaces in the cartilage from one stage to another until they became Haversian canals, still lined by the cells which he regarded as the descendants of the primitive cartilage corpuscles. He saw how the corpuscles of the Haversian canals became included within the bone to become its living denizens. He knew that similar cells built up the skeletons of invertebrate forms of life, and had no doubt that the human skeleton was laid down, not by arteries, but by an element of which Hunter had no conception—separate self-acting units of living matter. That discovery—Goodsir made it for himself, others had done the same independently—marks a new chapter in our knowledge of bone. But the old or Hunterian chapter did not pass out of date; the new chapter merely gave a more precise interpretation of the facts contained in the old. Bone corpuscles were substituted for arteries.

There were in the Museum of the Edinburgh College, under Goodsir's care, certain famous specimens which had been made in William Hunter's school in London by Cruikshank. They were supposed to demonstrate that lacteals opened on the intestinal villi by open mouths. Such openings or mouths were believed to exercise a selective function, picking out or absorbing from the aliment passing along the canal such elements as served for the nourishment of the body. Goodsir applied his microscope to these specimens; he could see the injected lacteals in the villi, but nowhere could he find openings or mouths. He applied himself to the problem of absorption. He fed a dog, and when digestion was at its height, examined the villi. He found cells laden with the products of digestion, and concluded that the carpet of epithelial cells which lines the bowel was the agent by which food was taken up and absorbed from the contents of the intestinal tract. His intimate knowledge of cellular processes in the lower forms of life led him farther; he realised that every cell in the human body must have a power of absorption, for every living cell must feed—must have the power of taking in nourishment. Absorption was a natural function of every kind of cell—including those which are found in bone. Thus he had to throw overboard Hunter's conception of lymphatic vessels playing the part of bone absorbers and bone modellers, and put in its place a very different conception—one of living units or cells which, from being builders, had become destroyers. When bone is absorbed it is not its lymphatics which had become thieves, as Hunter thought, but the cells living in its own interstices.

The microscope revolutionised, and at the same time simplified, our conception of the growth, maintenance, and reproduction of bone. In place of it being a living tissue manípulated by arteries and by lymphatics, it became an organisation of living units—bone corpuscles—each of them endowed with the power to build up or take down that territory which lay directly under its control. Goodsir was the first to realise the new picture of bone. That discovery has not made Hunter's assumption of "consciousness" in bone less tenable. How are such cells controlled? What are the circumstances which make them lay down bone on one day and carry it away on the next?

The investigations which Goodsir made on the nature of bone during the three years of his curatorship are recorded in four very brief papers which first appeared in his *Anatomical and Pathological Observations*, published in 1845.[1] One of these is entitled, "The Mode of Reproduction after Death of the Shaft of a Long Bone." It is an inquiry concerning the bone-forming power of periosteum—being an answer to Syme's investigations mentioned in an earlier part of this chapter. He concluded that periosteum has no bone-forming power—it

[1] See *The Anatomical Memoirs of John Goodsir*, by William Turner, vol. ii., 1868.

is simply, as he terms it, a "limiting membrane." He did not deny that periosteum "participated in the office of regeneration—an important principle in surgery," but its power to form bone depended wholly on the fact that particles of the true osseous substance of the shaft had become adherent to it, these particles becoming the centres from which the new bone attached to the periosteum was formed. By actual experiment he removed periosteum by scraping, and found particles of bone adhering to its deep surface. From such fragments, he concluded, new bone was reproduced. He examined the periosteal shells which form round necrosed shafts. He found that wherever there was a rough or ulcerated patch on the old shaft, there was also in the corresponding area of the periosteum a nucleus of bone growth; but where the areas of the old shaft were perfectly smooth—where there was no trace of exfoliation or ulceration of the old bone—there was a gap or cloaca in the overlying periosteal shell. If the periosteum owed all its osteogenetic power to adherent morsels of bone, then these appearances were just as they should be. He conceived that the shaft of a bone might be killed so quickly by disease that there was no time for the separation of living fragments to act as bone grafts on the periosteum. In such a case there would be no reproduction of bone. If the greater part of a shaft did die, he observed that ossific centres shot out most vigorously from the terminal living parts. He had never seen an epiphysis take any part in the regeneration of a shaft; in a physiological sense an epiphysis should be regarded as a separate bone. Finally, as regards Syme's experiments, Goodsir held that they did not demonstrate an osteogenetic power in periosteum, but they did demonstrate that, as detached by the surgeon, it could and did carry with it bone particles or grafts which could serve as centres of reproduction. For Goodsir, a bone was a hive of bone corpuscles wrapped within a nest of periosteum.

I have mentioned Hunter's graphic and accurate picture of what the unaided eye can see when a living bone casts off a dead piece. Goodsir's paper on "Absorption, Ulceration, and the Structures engaged in these Processes" adds a fitting pendant to Hunter's picture—a pendant painted through his microscope—the earliest picture of its kind. We have seen that Goodsir recognised the corpuscles which lined the walls of the Haversian canals as the oldest and most active of the cellular constituents of bone. He had observed them on the surface of the bone, under the periosteum. He knew well that these canalicular cells could be stimulated to activity by inflammation; in Hunter's phraseology, they were endowed with irritability. He saw, in an inflammatory area of bone, that these cells multiplied in the Haversian canals; the walls of the canal widened by the proliferation of tissue and the absorption of the wall until neighbouring Haversian systems almost met, or did fully meet. He observed that all along the margin which demarcated the dead from the living bone, there was this exaggeration of the normal activity of bone cells, accompanied by an increased vascularity, an expansion of the canals, and an absorption of the bony substance between the canals until an open trench was formed, which separated the dead from the living bone. The separation of a sequestrum was thus the work of osteoblasts. Goodsir had no difficulty in supposing that the cells which laid down bone could change their mode of life and devote themselves to clearing away their old handiwork. He conceived that the more active these bone-devouring cells were, the shorter would be the term of their lives, and the sooner would they melt and their remains pass into the general circulation. The microscope simplified the understanding of the effects of inflammation on bone: Goodsir was the first to realise that these changes were effected by living units or cells.

Goodsir's bone researches were all carried out in the three years of his curatorship. In 1846, at the age of 32, he was elected to the chair of Anatomy in the University of Edinburgh in succession to the third Monro. Just before the onset of his disastrous and long-drawn-out illness, some ten years later, he opened up a field of research which yet remains to be explored. It is one which is destined to throw a new light on the living qualities of bone corpuscles. In 1853 a pupil had sent him a new form of sponge dredged from the waters of Spitzbergen.

While investigating its structure, he observed that the spicules—its skeleton—had many forms, but each was built and laid down so as to strengthen and keep open the water-canals on the entirety of which the lives of the sponge colonists depend. He looked for the builders and found them. No one had seen or studied them before. He formulated the following law: "A spiculum is formed by a series of sponge-particles (sponge cells), so arranged in a column or system as, by the silification of more or less of their contiguous surfaces and substance, to form a continuous rod." In short, he saw groups of spicule-forming cells marshal themselves into a geometrical pattern and lay down supports of the exact form and in the exact position where such supports were needed. We shall see that bone corpuscles have also similar qualities.

During the last ten years, Professor Dendy,[1] of King's College, London, and his former pupil, Professor Woodland,[2] have carried out a series of researches which justifies one in regarding the scleroblasts which form the spicular skeletons of invertebrate forms of life as the exact counterparts of the osteoblasts of our own bodies. In the more primitive types a spicule is actually formed inside the cell—an intracellular secretion; all stages between the intracellular and extracellular mode of spicular formation can be demonstrated. The scleroblasts are special cells derived from the dermal or skin layer. They do their work according to the needs of the animal frame of which they form part; they build their spicules so as to strengthen that side of a sponge which is exposed to the force of the waves, just as osteoblasts built to withstand the pressures applied to bone. The scleroblasts, Professor Dendy observed, are sensitive to vibrations. Thus we shall see that at an early stage in the evolution of animal life very wonderful cells came into existence, cells which could design and build struts and supports to meet the most varying forms of stress and pressure. The scleroblasts, the "mason-cells," of sponges, to which we have merely alluded, work in silica, but others build with calcium

[1] For Professor Dendy's latest paper, see *Proc. Roy. Soc.*, 1917, lxxxix. 573.

[2] *Quart. Jour. Microscop. Sci.*, 1905, xlix. 231; 1906, xlix. 533.

salts, just as do bone cells. We must suppose that the cells which build the skeletons of vertebrate animals are the descendants of such cells and have inherited all their marvellous aptitudes. But while we can study the behaviour of osteoblasts only under the most obscure and difficult circumstances, their ancient cousins lie open to us, clear and diagrammatic in their outlines and easy of access to the prying eye of the microscopic lens. It was John Goodsir who pointed the way to this lowly source of knowledge. We have not yet said good-bye to him; he again becomes our guide in several of the chapters which are to follow.

Louis Xavier Edouard Leopold Ollier,
Professor of Clinical Surgery in the University of Lyons,
Born 1830; Died 1900.

CHAPTER XVI

RESEARCHES INTO BONE GROWTH AND BONE REPRODUCTION BY OLLIER OF LYONS AND MACEWEN OF GLASGOW

In this chapter we shall examine the discoveries and researches which form the basis of our modern knowledge of bone growth, bone repair, and bone reproduction. In the modern period we shall find that surgeons are still divided into "Duhamelites" and "Hallerites." Louis Ollier, who was born near Lyons in 1830 and afterwards served as Professor of Clinical Surgery and surgeon to the Hôtel-Dieu of that city, became a convinced supporter of the Duhamel theory that periosteum is the maternal tissue of bone. That conviction was founded on a prolonged period of experimental investigation (from 1857–1868), and of clinical observation. He died in 1900 in his seventieth year. Sir William Macewen, born in the island of Bute in 1848, studied Medicine in the University of Glasgow when Lister occupied the chair of Surgery there—the chair which he also was called on to fill in 1892. His experimental investigations and clinical observations led him to a conclusion which is exactly the opposite of that reached by Ollier—

namely, that periosteum has no power to form bone. Flourens, whose researches form an essential link between the investigations of Duhamel and those of Ollier, was born in the Department of Herault in the south of France in 1794, and died in 1867. At the time he carried out his researches on bone growth he held the chair of Comparative Physiology in the Natural History Museum, Paris. He was a supporter of Duhamel. In another chapter we have had an opportunity of studying this brilliant Frenchman.

IN almost the selfsame year that Goodsir of Edinburgh was quietly establishing to his own satisfaction that bone corpuscles were the agents which built bones up and took them down, Marie Jean-Pierre Flourens, the permanent secretary of the Academy of Science of Paris, was proving, not only to his own satisfaction, but to that of all the world, that Duhamel was right: periosteum was the maternal tissue of bone. When he published his *Recherches sur le Dêveloppement des Os et des Dents* in 1842, he was a man of 48, with behind him the most brilliant record of experimental success of any man in Europe. He had that gift which is often given to Frenchmen —a clear incisive engaging style which carries conviction with every sentence. He had repeated and extended Duhamel's observations. He agreed that Duhamel was right: new bone was deposited on the surface of the bone under and by the periosteum. He also agreed that Hunter was right: absorption of bone was continually taking place; the substance of every living bone was in a constant state of flux. He had destroyed, as Troja had done, the medulla and medullary lining of a long bone, and seen, as the result of the experimental injury, the whole shaft die and the periosteum start into active life and lay down a new shaft. Troja's experiment he regarded as positive proof that the periosteum could reproduce bone after the entire substance of the shaft had been destroyed. He emphasised the fact that growth in length took place at the epiphyseal lines. People knew of the discoveries of Flourens; those of Goodsir were overlooked.

In 1857 Flourens was ageing rapidly, when a young medical graduate of Montpellier, of the name of Louis Ollier, came to Paris, attracted above all by the great name of Claude Bernard, the genius of experimental research. Ollier was a son of Southern France; although a graduate of Montpellier, his real university was Lyons; there he had studied under Bonnet, the leading authority on diseases of the joints when Ollier was studying Medicine. In his student days the functions of the periosteum had been debated; Ollier held the opinion that it had no osteogenetic function—in spite of the researches of Duhamel and Flourens. During the short period which he spent in Paris he determined to settle the matter for himself by applying the experimental method of Claude Bernard. He went to work in the laboratory of Chauveau. He selected a young rabbit as the subject of his first experiment. He lifted from the subcutaneous surface of the tibia a tongue-shaped flap of periosteum, leaving it attached by its base. He turned the flap or strip over the deep flexor muscles of the leg so that it formed a kind of annular ligament over them. In six weeks he found that the membrane had become a loop of bone, thick and strong at its base, but tapering towards the tip until it became fibrous. He repeated the experiment in a slightly different form and got the same result; a semi-detached piece of periosteum could and did form bone. He then proceeded to vary the conditions of his experiment further. He twisted the base of the periosteal flap so that the vessels entering at its base would be compressed; the twisted slip of periosteum produced a beaded string of bony nodules. In a further experiment he raised a process of periosteum, turning it, as before, over the adjoining muscles. Three days later he detached its base from the shaft of the tibia. The slip of periosteum he had thus detached also produced bone; the power of producing bone, he concluded, must be inherent in the periosteum, for after the base of the flap was cut there could have been no migration of osteoblasts from the parent bone to the detached slip. He then raised, and at the same time completely detached, a piece of periosteum, and transplanted it into the subcutaneous tissue of the thigh. He used, for the first time, periosteum as a graft; that graft produced bone. Ollier entered Chauveau's laboratory in Paris a sceptic; when he had completed his first series of experiments he was convinced of the osteogenetic properties of periosteum.

We must look closely at Ollier, because he is almost the

earliest example we shall meet of a new type of surgeon—one who realised that experimental physiology is, and must be, the basis of all rational operations in surgery. Lister and he were almost contemporaries: Ollier was born in 1830, three years after Lister. Lister gave the first ten years of his career to experimental physiology; so did Ollier. The series of experiments we have just narrated—done when Ollier was in his twenty-eighth year—would have satisfied most young men. Ollier had the qualities of the true investigator: he realised that his experiments had touched only the fringe of the physiology of bone. Hence, when he returned to Lyons in 1859, where he was soon to be called to fill the chair of Clinical Surgery and serve as surgeon to the Hôtel-Dieu, he continued his investigations year after year until in 1868 he published the results of his experimental and clinical observations in two volumes under the title of *Traité Expérimental et Clinique de la Régénération des Os et de la Production Artificielle du Tisssu Osseux*. From the time he obtained his first experimental results Ollier had one major object in view—namely, to demonstrate that the reproduction of a joint, which had been excised because of disease, was only possible if the surgeon conducted the operation so that the periosteum was left intact. He was convinced that the terminal ends of bone could be reproduced only by the periosteum.

When he commenced a second series of experiments in 1858, he devoted his attention more closely to the anatomy and microscopic structure of the periosteum. His critics had alleged that when he raised the periosteum—which he did by pressing a rugine firmly against the surface of the shaft of the bone—he also detached fragments of bony matter, and it was these and not the periosteum which gave rise to new bone. He therefore examined with a lens his detached strips of periosteum, and satisfied himself that they were free from visible fragments of bone. He was a proficient microscopist, and studied the minute structure of the periosteum. He defined that membrane as including all the soft tissue which covers the surface of a living bone. When a section of periosteum was studied while still attached to the surface of a growing bone, he could observe a gradual transition of structure from its superficial to its deep surface. As its deepest stratum approached the bone, he observed that it became more cellular in nature, and where it actually mingled with the bone, that these cells were clearly of the bone-forming type—they were osteoblasts. Osteoblasts, he found, were present in the deepest stratum of the periosteum; its deepest layer formed an osteogenetic tissue. It was true, he said, that these osteoblastic cells were sparsely scattered in the deepest stratum of the periosteum of an adult animal, but he also demonstrated that they could be quickly increased in number by injuring and thus irritating a bone. He had discovered, as Goodsir had done, that the cells lining the canaliculi of bone rapidly increased in number and in activity when the parent shaft was injured. They were roused to activity much as bees in winter when their hive is tapped.

So he again resorted to his "flap" experiments. He scraped the deep surface of such tongue-shaped strips of periosteum with his knife, and the soft juicy matter thus collected was transplanted to a living subcutaneous bed; from the matter thus transplanted fine granules of bone were formed. He had thus proved—at least he thought he had—that the deepest stratum of the periosteum had an osteogenetic function. Raising the tongue-shaped flaps from the periosteum of the tibia, he destroyed the deep or osteogenetic layer of one part of it by scraping or by applying a cauterising fluid, while in the remaining area of the flap the deep layer was left uninjured. On turning such a flap over a muscular bed, and leaving it in that position for a period, he observed that only the intact or uninjured area produced bone. That was, in Ollier's estimation, another proof that the deep stratum of the periosteum produced bone.

Then followed a series of experiments in which he completely detached flaps of periosteum and transplanted them as grafts—heterotopic grafts, he called such slips—to other parts of the body, where they held and produced bone. He found that the success of such grafts depended on many circumstances. There was, in the first place, the occurrence of suppuration.

Suppuration, he found, could kill part of a graft and leave another part unaffected ; much of Ollier's work is invalidated because he did not know the cause and effects of suppuration. Grafts from young animals took more readily than those from adult animals. Those from the shafts of long bones held better than those from flat bones. He observed that periosteal grafts of rabbits had a greater power of reproduction than those of cats. The power, he found, varied with the bone : a periosteal graft from the nasal bone of the cat would hold, while that from the corresponding bone of the rabbit would fail. He showed that the dura mater had a greater osteogenetic power, when transplanted, than the pericranium. All of these results represent real additions to our knowledge of bone life.

Having thus completed his researches on the periosteum, and satisfied himself that it possessed an osteogenetic power to a high degree, Ollier turned his attention to the marrow. He tied the nutrient artery to see if the main arterial supply was necessary for the life and sustenance of the marrow ; he found that ligature made no apparent difference because marrow has many subsidiary sources of blood supply. He failed to raise grafts of bone from transplanted marrow. But he observed that he could increase its osteogenetic power by irritating the bone by injury. To test its osteogenetic power in another way, he introduced a new kind of experiment. He amputated the foot and lower part of the leg of a rabbit, and into the medullary cavity of the tibia thus exposed he introduced an open metallic tube. At the end of a month he found that the tube was filled with ossifying granulation tissue ; that tissue, he concluded, was produced by the marrow, and therefore marrow did possess an osteogenetic power in a low degree.

Having thus tested the powers of periosteum and marrow to form bone, he now passed to bone itself. He found that a piece of bone, detached from periosteum and medulla, could not serve as a graft ; it died, and ultimately was absorbed. In his opinion a bone-graft acted amongst living tissues as a foreign body. One would be inclined to suppose that Ollier's experiments on the transplantation of bone fragments had been vitiated by sepsis, did he not go on to describe experiments where such grafts held when they were transplanted with periosteum and marrow still attached to them. In his opinion, pieces of living bone with their periosteum and marrow still intact provided the most perfect form of bone graft; he supposed that the periosteum served as an intermediary between the bone fragment and the muscle in which it had been implanted. Yet Ollier knew that the shaft of the tibia of a young animal could be completely denuded of periosteum and its marrow entirely removed, and yet live and grow. The results thus obtained confirmed his belief that periosteum was the chief bone-producing tissue.

Ollier made experimental fractures; the conclusions he drew from them were: (1) The periosteum is the chief agent in forming the callus; (2) The marrow takes a slight part; (3) The bone itself takes the least share. These were almost identical to the conclusions reached by Duhamel and by Flourens.

He opened up another new subject when he began to study the effects of experiment on epiphyseal lines of growth. He observed that irritation of the shaft of a bone, by injury, could increase the rate of growth at these lines. Such a result tells one as plainly as can be that there is a mechanism in every bone for correlating the activity of its army of living cells—bone-laying and blood-carrying cells. He discovered that the epiphyseal disc of cartilage could keep up a continuous supply of material for increasing the length of a bone only when it formed an intrinsic part of that bone. When the disc was cut out and transplanted, its growth failed; it became a disc of bone. It was known before Ollier's time that in every long bone there is an epiphyseal disc of "maximum growth"—a disc at which the major part of the length of a shaft is produced. The maximal discs for the femur and tibia lie respectively above and below the knee-joint, whereas those for the humerus and radius are placed at the shoulder and at the wrist. Ollier was the first to study experimentally the exact shares taken by the proximal and distal epiphyses in producing the growth of a bone in length.

It is unnecessary here to enter into three other series of

experiments carried out by Ollier. He repeated Duhamel's experiments of boring holes in the shafts of growing bones and found, as Duhamel had done before him, that there was no interstitial growth—that growth in length took place only at the epiphyseal lines. In another series he repeated, and extended in great variety, the kind of experiment carried out by Syme—excision of the shafts of long bones, or parts of a shaft, to discover the share taken by the periosteum in the reproduction of new bone. His conclusion was that the periosteum was the all-important factor; the surgeon must preserve it when he performed a resection if he would hope for a reproduction of the bony parts which it was proposed to excise. Then there was a third, and in Ollier's opinion the most important, series—those in which he studied the reproduction of joints. His main conclusion one could guess—resections of joints must be performed so that the periosteum covering the excised parts was preserved. He found cartilage to be the most inert, the least endowed with the powers of vitality and reproduction, of all the tissues of the body. On only one occasion did he find a transplanted piece of epiphyseal cartilage produce bone; he never did see articular cartilage reproduced within a grafted articulation.

So far we have been following Ollier as an experimental physiologist; we have space merely to glance at one of his clinical cases—the shoulder-scarred little maid so beautifully portrayed in *Plate VIII.* of Vol. II. of his treatise. Her name was Louise Gaillard; she came under Ollier's care in September 1864, a meagre tuberculous child of five, with suppuration of the shoulder and arm. Ollier removed the upper half of the diseased but living humerus—head, neck, and shaft—by his subperiosteal method. The bone was reproduced which ultimately became a facsimile of the one he had excised. In the following year, 1865, Louise was able to work in a cigarette factory, and a year later, when she was seven, the restored humerus was only 46 mm. shorter than its normal companion. This is one of the classical cases of resection of the humerus. We have noted several others—Charles White's case of 1768, and Syme's case of 1826, and there is another, a resection of the humerus by Hunter. There is also a fifth and yet more famous case which we shall examine in a subsequent part of this chapter.

We shall see presently that Ollier made a grave blunder in the course of his experiments. But when we have made that admission, we have to recognise him as the man who established the evidence on which the principles of bone surgery must be founded. It was he who gave orthopædic surgeons materials for framing a rational charter. The total result of his ten years' investigations was to make him a confirmed Duhamelite—a worshipper of periosteum. We take our leave of Ollier of Lyons with the completion of his great *Treatise* in 1868; he was destined to see the commencement of the twentieth century and his labours fully recognised by a new generation of surgeons.

We are now to return to Glasgow. The occasion of our former visit to that city was in 1877, when William Macewen, then aged 29, introduced a new operation for the cure of knock-knee. At the same time he was rectifying the bent tibiæ of rickety children by the excision of wedge-shaped pieces. In the following year, 1878, a boy, three years of age, was brought to him at the Glasgow Infirmary, emaciated and exhausted from suppuration following acute osteomyelitis of the right humerus. Almost the entire shaft of the bone was dead; after resection had been performed, there remained in the boy's arm only the extremities of the humerus, the lower being represented merely by the epiphysis, while the upper was made up of the epiphysis and a small part of the shaft. In 1880 the boy was brought back by his parents to have his arm amputated; it was useless, although the muscles were healthy and could act. No part of the humeral shaft had been reproduced. Making a longitudinal incision deep into the outer aspect of the arm, Macewen planted along the muscular furrow thus opened out a row of tibial grafts. They were pieces obtained by fracturing wedges which had been cut from the bent tibiæ of various boys. Most of these grafts were devoid of periosteum, yet they held and lived. As thus reconstructed, the boy's humerus was six inches in length, being made up of three elements: $4\frac{1}{4}$ in. (10·7 cm.) of the shaft, composed of tibial grafts; the upper part, $1\frac{3}{4}$ in. (4·3 cm.), derived from the original upper extremity; and there was also the persistent lower epiphysis. In 1910, when the boy had become a capable workman of 35, he had a useful, well-modelled, if curved, humerus, which measured 11 inches, being 3 inches shorter than the normal bone o the left arm. If Ollier's researches had revealed the ful truth, then those broken tibial fragments ought to have died and been absorbed; and yet there can be no doubt they had lived, and given rise to a shaft which had become connected with the persistent upper and lower extremities, and thus become parts of a functional humerus. When he sowed the grafts, Macewen could not have missed noting any ossifications which had survived from the old periosteum. No such ossifications were present. By this operation Macewen demonstrated, for the first time, that fragments of bone may be used as grafts, and that such grafts will live, grow, and reproduce bone. Further, that the bone reproduced by such grafts can, under tensions and stresses brought to bear on them by the muscles of the part, be moulded to form a real humerus—real in an anatomical as well as in a physiological sense. Macewen's case forms the fifth and greatest of a classical group of famous clinical humeri.

At the time when Macewen was employing bone grafts with great success in Glasgow, Ollier's opinions were in the ascendant in the rest of the world. At first the significance of Macewen's discovery and practice was not generally appreciated, nor was it later. Hence, fully thirty years after he had begun to use fragments of bone as grafts, Macewen, in order to force the truth of his discovery on the attention of surgeons, carried out a series of operations on animals.[1] He repeated Syme's experiments; he cut out, subperiosteally, $1\frac{3}{4}$ inch from a young dog's radius. At the end of four weeks there was still a gap in the bone; the periosteum, although it had been carefully preserved, had failed to form the missing part of the shaft. He then, in another animal,

[1] *The Growth of Bone; Observations on Osteogenesis*, Glasgow, 1912.

cut out the whole radius, leaving the periosteum in place; the periosteum again failed to reproduce bone. He carried out modifications of Ollier's experiments, raising flaps of periosteum from a young dog's radius, folding it round adjacent muscles of the foreleg. Both of his Ollierian flaps of periosteum failed to produce bone. He transplanted grafts of periosteum as Ollier had done, yet failed to succeed in raising bone from them. At first sight it almost looks as if the dogs of Edinburgh and Lyons were provided with a different kind of periosteum to that which clothed the bones of dogs in Glasgow—in such sharp contrast stand the results gained by experimenters in these cities. Yet all three surgeons are equally credible witnesses. There is only one explanation of their discrepant results; Syme and Ollier must have removed the soft tissues which cover bone to a greater depth than did Macewen. The evidence that periosteum, as usually raised by surgeons from the surface of the bones of a growing animal, *does produce bone* is so positive, so unassailable, that it cannot be rejected because a certain series of experiments gave negative results.

In order to study the effect of periosteum on the growth and life of a bone, Macewen removed this membrane from the radius of a young dog. He found, as Ollier had done, that the bone lived and that its growth was slightly retarded. The shaft lost its smooth contour. On such denuded bone he applied modifications of Duhamel's ring experiment; the rings of silver wire with which he encircled the bone became embedded in the shaft although the periosteum had been cut away. It was clear, from these experiments, that new bone could be formed on the surface of a shaft—the shaft could continue to grow in thickness after the periosteum was cut away.

Sir William Macewen carried out a series of experiments of a new kind to substantiate his clinical observations—namely, that bone itself could be grafted, and that the graft could produce new bone, independently of periosteum. He took two dogs, cut an inch out of a radius of each, removed the periosteum from the pieces excised, and broke each of these into fragments. He reimplanted the pieces so that the two dogs exchanged their radial fragments. The grafts held, and complete restoration of the radial shafts resulted. He took two other dogs, and exchanged the right radius of the one for the corresponding bone of the other. In each case the radius was stripped of its periosteum; yet the transposed bones lived, and became thicker than their untransplanted counterparts. He found that bone shavings, but not bone dust, served well as grafts; the greater the superficies of a fragment, the better he found it to serve as a graft. The greater the surface of the graft, the greater the area for nourishment and reproduction. Macewen demonstrated the practical importance of his discoveries by sowing fragments of living bone in deficiencies in the cranial vault. There the grafts held and filled up the cranial breach.

One cannot but admire Macewen's resource in devising experiments. We have seen that Ollier inserted metallic tubes within the medullary cavity to test the power of marrow to reproduce bone. Macewen used glass tubes, not to test the growth of marrow, but to exclude the periosteum from taking any part in filling up the gap left in a bone when a piece of its shaft had been excised. Having cut out a segment, he inserted in its place an open glass tube, so that the open ends of the tube faced the cut ends of the shaft. The tube became filled from the shaft with growing tissue which ossified. The new bone thus formed within the tube must have come from the shaft; it could not have come from the periosteum which was excluded by the wall of the tube. He conceived it possible that an absorbable ferrule placed round the adjacent ends of a broken bone might acclerate union by serving to confine the reparative bone-forming mass formed by the cut ends of the shaft. Such a tube took the place of the periosteum, which Macewen regarded as a "limiting" membrane.

We do not suppose that the briefly-narrated researches of John Goodsir to which we have alluded in a former chapter, were known to Sir William Macewen when he carried out these experiments; yet his definition of the function of the periosteum, as a membrane to limit the excursions of the osteoblast and to prevent bone-building cells from invading surrounding tissues, is an exact replica of John Goodsir's conception. The observations which Goodsir made in the dissecting room and his laboratory during the fifth decade of the nineteenth century led him to formulate exactly the same conception of bone growth as Macewen arrived at seventy years later from a life of active clinical and experimental observation. Macewen, however, has much better evidence of the limiting function of the periosteum to lay before us than had Goodsir. He cites the evidence from a case where osteoblasts had been carried by the blood within an aneurysmal sac where they had set up ossification within its fibrinous walls; he describes cases where the exudate escaping from between the fractured ends of a femoral shaft had been disseminated by massage into the surrounding ruptured muscles, where it formed a mass of ossific tissue; he cites instances of fractures where the periosteum has been much torn, thus permitting osteoblasts to escape into the surrounding tissues and form bone. He alludes to cases where the periosteum has fallen between the broken ends of bones and prevented union. In such cases he supposed that the periosteum served as a barrier, as a limiting membrane, between armies of osteoblasts which advanced to repair the fracture. Macewen thus gives a logical explanation of Hunter's observation that the periosteum or tissues which surround a broken or diseased bone may take on an "ossific disposition."

There is another series of Macewen's experiments to which we must allude before we leave his work. We have seen that Ollier studied the epiphyseal discs of growth. Macewen rightly protests that such discs have nothing to do with epiphyses; the epiphysis has its own peculiar independent nature. The growth disc is part of the shaft—of the diaphysis —and therefore ought to be called the diaphyseal disc, or shaft-disc. It is solely concerned, as he has demonstrated, with the growth of the shaft. He removed the shaft of the radius from a young dog, leaving merely the epiphyses and diaphyseal discs. In six weeks the shaft was reproduced, not from the periosteum, but from the diaphyseal discs. The point of meeting between the parts of the shaft produced from the proximal and distal discs could still be seen in the restored radius. To make certain that his interpretation was right, he removed $2\frac{1}{2}$ inches from the shaft of a growing dog's radius and covered each cut end with a metal cap. In seven weeks the caps had met, pushed together by the increase at the diaphyseal lines. He removed the distal third of the shaft of a radius, including the epiphysis ; he found, as Ollier had done, that the part excised was reproduced, the new distal part of the radius being provided with an epiphysis, but there was some degree of shortening. Ollier believed that in such cases reproduction was effected from the periosteum ; Macewen has demonstrated that there is a downward growth of the shaft itself, which replaces the part excised. The idea that compensatory growth at the diaphyseal discs may be utilised by surgeons to make good an extensive gap in the shaft of a bone is a discovery we owe to Macewen.

Macewen's work brings us down to the position of our knowledge regarding the growth and reproduction of bone at the present day. We have seen Duhamelites and Hallerites come into existence in the eighteenth century; we have seen them in the persons of Syme and Goodsir in the nineteenth century ; and here we are in the twentieth century still with the same two sects—those who regard periosteum as the chief osteogenetic element of the human skeleton, and those who regard bone itself as that element. Where is the truth ? As in most cases of this kind, there is truth on both sides, but the greater share is on Macewen's. My interpretation of his work is this. Orthopædic surgeons lay it down as a basal principle that a deformity, to be remedied, has first to be over-corrected. To remove an error one has to exaggerate the truth, and to compel surgeons to listen to the truth about bone its virtues had to be enhanced. That is what I think Macewen has done. His has been a great practical service; he has given bone its rightful place among the living tissues of the body—a tissue which has the power to reproduce itself. But in raising bone to its proper biological and surgical status, he has been less than just to periosteum. Periosteum can reproduce bone, for no one has yet made a microscopical examination of the periosteum of growing or of inflamed bone and failed to find in it those very elements which Macewen admits to be the creators of bone-osteoblasts. From an anatomical, and I think also from a surgical, point of view, the osteogenetic power of the deepest stratum of the periosteum cannot be denied.

CHAPTER XVII

THE INTRODUCTION OF THE MODERN PRACTICE OF BONE GRAFTING

In this chapter a brief account is given of Hunter's experiments in grafting of animal tissues. A résumé is given of the researches by Ollier of Lyons, and of the modern use of bone grafts to serve as a means of uniting the fractured ends of broken bones.

THE history of human invention is ever the same: we find that the path which led to final success is always strewn with the wreckage of efforts which failed. We often wonder why these efforts failed; looking back, with the knowledge of the present, we can see they often had in them all the elements of success, and yet they failed. Why did John Hunter not succeed in giving the practice of grafting, which has at last gained the recognition of surgeons, a permanent place in surgery? He had come as near as any man has yet attained to an understanding of the nature of living matter and of the conditions which are necessary for the successful implantation of a living graft amidst living tissues. We see he failed because of sepsis; sepsis ruined the brilliant programme he had conceived. If Pasteur and Lister had been born before him he would have succeeded and been saved quests which he pursued in vain. Many of Hunter's projects failed because they appeared before the times were ripe for their reception.

In a former chapter we have seen how Ollier of Lyons, in 1858, sixty-five years after Hunter was dead, and when his grafting experiments had passed out of mind, commenced a systematic series of attempts to transplant periosteal grafts. He succeeded and yet he failed; his experiments did not induce surgeons to employ periosteal grafts to remedy defects in bone. His work fell dead from his laboratory on account of two reasons: (1) Because of sepsis; (2) because he did not realise that certain conditions are necessary to stimulate bone-forming cells into activity. Bone and muscle cells are alike in this: their growth is dependent on action; they thrive only so long as they have to work. Ollier wondered why the fragments of bone which he had succeeded in raising from slips of periosteum planted beneath the scalp or amongst muscles, ceased to grow and tended to disappear. These bony grafts withered because they were not subjected to the strains and stresses which rouse the activity of osteoblasts. To use a Hunterian phrase, they were "conscious of imperfection" and hence became absorbed.

The modern practice of bone grafting was invented by Macewen in the Infirmary of Glasgow in 1880. The conditions which determined his success were these: (1) He operated in the building where, fourteen years previously, Lister had shown how sepsis could be conquered. (2) He held the opinion that bone was, as Hunter and as Goodsir had thought, a living tissue which possessed the power of surviving, reproducing itself, and of thus being able to serve as a graft. (3) By a fortunate chance he planted his tibial grafts in a situation where they soon became subjected to muscular stresses and strains. In a short time bony fragments gathered from the legs of six boys became intrinsic parts of the humerus of a seventh; from the moment of primary union the bone cells of the grafts were brought under the stimulating impulses of the biceps and triceps. Osteoblasts are the obedient slaves of muscles; muscular dominance is their breath of life. The reconstruction of that boy's right humerus in Glasgow Infirmary in 1880 is the first paragraph of a new chapter in the history of surgery. The paragraph could not have been written earlier, because the necessary data were not in existence for its composition.

The impulse, however, which has given bone grafting its present extended use in orthopædic surgery came not from Sir William Macewen's initial success, but from another operation altogether. I have described in a former chapter the conditions which induced Arbuthnot Lane to introduce operative means for the treatment of fractures—particularly for those of the leg. He really introduced a new principle—the principle of internal splints. Until April 1894 surgeons sought to maintain apposition and immobilisation of fragments by means of external splints. Arbuthnot Lane conceived the idea of internal splints. His internal splints were metallic. We shall see that modern American surgeons have substituted for Lane's screws and plates, live bone grafts to serve as internal splints.

When I now turn backwards to examine the experiments carried out by Hunter on grafting, it is not to extol his knowledge or his prescience: that would be a fruitless task. My object is much more utilitarian; it is simply to bring before you certain neglected observations of his, which are of service in surgery, not only at the present time, but for all time. In the strict sense of the term Hunter never did graft bone. It is true he watched those wonderful masses of bone which Nature grafts and rears each summer on the heads of deer—studied their growth, decay, and death. Hunter's knowledge of bone grafting is founded on his experience of tooth and of spur transplantation. He rightly regarded dentine as a modified form of bone. The root of a tooth is covered by bone; it is secured to its socket by periosteum; the conditions of successful tooth transplantation are those which regulate success in bone grafting. We therefore turn for a moment to Hunter's experience in this department of surgery.

Let me first quote a case from his practice. "A gentleman had his first bicuspis knocked out and the second loosened. The first was driven quite into his mouth, and he spat it out upon the ground, but immediately picked it up and put it in his pocket. Some hours afterwards he called upon me, mentioned the accident, and showed me the tooth. It was not quite dry, but very dirty, having dropped on the ground and having been some time in his pocket. I immediately put it into warm water, let it stay there to soften, washed it as clean as possible, and then replaced it, first having introduced a probe into the socket to break down the coagulated blood which filled it. I then tied these two teeth (the detached first and the second loose, premolar) to the first grinder and to the canine with silk, which was kept on for some days and then removed. After a month they were fast as any teeth in his head." Hunter replaced that tooth because he knew it was still alive; he knew it could be kept alive for at least twenty-four hours outside the body. The success of the operation of transplantation, he writes, "is founded on a disposition in all living substances to unite when brought into contact with one another."[1] He discovered that dead teeth could be implanted; in some cases they held, even cleared up so as "to take on the appearance of living teeth"; but in most cases they did not hold, or if they held, their roots became eroded and absorbed, and such grafts dropped out.

Let us look for a moment at the criteria applied by Hunter to measure the vitality of a part. He regarded enamel as a dead substance (1) because it had no power of repair; (2) it contained so slight an amount of animal matter; (3) because madder, when circulating in the blood, never stained the enamel of teeth. The dentine, or, as he called it, the "bony matter" of a tooth, although it possessed no blood vessels and thus lay outside the blood circulation, had "most certainly a living principle." His belief was founded on the following observations: (1) That dentine could react; as the crown wore down, new dentine was deposited on the roof of the pulp cavity. (2) It contained a larger percentage of "animal matter." (3) If a growing animal were fed on madder, the dentine, then being formed, combined with the madder circulating in the blood; dead bone did not become stained. He inferred that the vitality—or, to use our modern terminology, the metabolism—of dentine was low, because the red stain was permanent. In bone the stain was not permanent; in time the madder-coloured parts were absorbed. We see then that Hunter had realised that a part could live outside the circulation; it could live when nourished only by the lymph which exuded from the blood vessels. Having reached that belief early in his career, it was easy for him at a later stage to

[1] *Collected Works*, vol. ii. p. 161.

realise that a graft of any kind, when transplanted, could maintain life from the juices of the part long enough to await the period of vascularisation. Hunter, it must be remembered, had proved that the umbilical artery was still alive three days after the placenta had been shed.[1]

He lays down the conditions for successful transplantation. The first is that the tooth-socket must be fresh, healthy, and, above all, free from disease. Grafts from young animals, or from those low in the scale of life, were more vigorous than those taken from adult animals or from those high in the scale. "For the living principle in young animals and those of simple construction . . . continues longer in a part separated from their bodies, and even would appear to be generated in it for some time, while a part separated from an older or more perfect animal dies sooner." The operation of transplantation had to be done "in such a way that she (Nature) can assist."

I will not stop now to describe the museum preparation which shows a human tooth engrafted when it was alive in the comb of a cock. There is a vascular union between the comb and the tooth, but there is no proof that the tooth actually lived in its new situation, although Hunter believed it to have lived. I want to pass on to his spur transplantations. A spur, as you know, has a core of solid bone—an outgrowth from the tarso-metatarsus of the fowl's leg. Experiments on spurs are experiments on bone. Hunter wished to discover if one sex was superior to the other in reparative power:

"I took the spur from the leg of a young cock, and placed it in the situation of the spur in the leg of a hen chicken; it took root; the chicken grew to a hen, but at first no spur grew. This experiment I repeated several times in the same summer, with the same effects, which led me to conceive that the spur of a cock would not grow on a hen, and they were, therefore, to be considered as having distinct powers. In order to ascertain this, I took the spurs of hen chickens and placed them on the legs of young cocks. I found that those which took root grew nearly as fast and to as large a size as the natural spur on the other leg." How obtuse John Hunter

[1] *Collected Works*, vol. iii. p. 159.

can be at times. He knew that the testes could by a "species of sympathy" completely alter the form of an animal's body, and yet when he came to draw the lesson which those classical experiments of his have to teach us, he lamely says: "The spurs of a cock were found to possess powers beyond those of a hen, while at the same time the one animal as a whole has more power than the other; yet when I apply these principles to the power of cure of local diseases of the two sexes in the human race, I can hardly say I have observed any difference."[1] Those who have noted the results obtained by Hunter in the transplantation of spurs, and have marked the changes which occur in acromegaly, see in them the effects produced by substances which are set free by the testicle and by the pituitary body. There are substances or hormones which can sensitise —can activate bone, and could therefore, if our knowledge were more complete, be brought to bear on bone grafts. In reality those spur experiments of Hunter are amongst the most instructive in the whole range of the literature on bone grafting.

Hunter's researches take us back to the eighteenth century; we are now to glance at the practice of bone grafting as it is being carried out by surgeons in America in this the second decade of the twentieth century. We shall commence with a neatly conceived series of experiments carried out by Dr. W. E. Gallie of Toronto in 1914.[2] He reflected the periosteum from a small segment of the outer border of the radius of a dog; with a saw he separated a small wedge of the bone; having separated the wedge, he immediately replaced it in the shaft and stitched the periosteum over it. He performed this operation on three dogs, killing them, so as to obtain bone grafts at three stages of implantation—at the end of the first, the second, and the third weeks. The result of examining these grafts was to lead him to the conclusion, as similar experiments had led many men before him, that a bone graft dies and merely serves as a scaffolding, which is invaded by the neighbouring living bone. He performed three other

[1] *Collected Works*, vol. iii. p. 274.

[2] *Journ. Amer. Orthop. Surg.*, 1914, vol. xii. p. 201.

modified wedge experiments. In one he boiled and killed the wedge graft before replacing it; in another he replaced the wedge taken from a dog's radius by one taken from a cat's; in a third he surrounded one side of the cut surfaces of the wedge graft with tinfoil. The results yielded by the second series of experiments confirmed the conclusion he drew from the first series, namely, that a bone graft merely supplies a favourable nidus for the invasion of the host bone. We could cite numerous references to experiments which have led their authors to similar conclusions.

Yet, conclusive as such experiments seem to be, there lies somewhere in them a fallacy. They are at the best negative experiments, yielding negative evidence. If Macewen's tibial grafts had merely served as scaffolding for the living bone of the distant extremities of that Glasgow boy's humerus, could the shaft of the bone have been reproduced in the short time which was taken? Albee has published excellent radiograms of a fibula, which was accidentally fractured soon after being transplanted to replace a necrosed tibia. The fracture occurred near the middle of the shaft, far distant from the implanted extremity at which an invasion of tibial osteoblasts would take place. At the site of the fracture a healing process immediately set in; a callus was formed by the graft, which we therefore must suppose to have lived through its whole extent. We have other proofs in the behaviour of transplanted fibulæ; their appearance, as revealed by X-rays, although they show thickening to occur first at the implanted extremities, supports the idea that grafts can retain living osteoblasts in their whole extent: from what we know of the anatomy and physiology of bone, we should marvel if its cells, when suitably transplanted, should fail to live and grow. The evidence as it now stands leads one to suppose that in every fragment of bone successfully grafted, at least one colony of osteoblasts remains alive to serve as a centre of regeneration.

In 1911, Dr. Fred. H. Albee,[1] Professor of Orthopædic Surgery in New York Post-Graduate Medical School, initiated a new practice — that of employing living bone grafts as

[1] *Bone Graft Surgery*, New York, 1915.

internal splints. To fashion splint grafts of the exact shape and size needed for such operations, and to obtain them with a minimum of labour, he introduced from the cabinetmaker's workshop to the operating theatre motor-driven cutting machines of the latest design; with a twin circular saw he found it possible to cut from the shaft of a tibia miniature planks of living bone, contriving his sections so as to have periosteum on one side and a layer of medullary lining on the other. The first use he made of such grafts was to immobilise the damaged region of the spine in cases of Pott's disease. At the site of an incipient kyphosis he exposed the spinous processes, split them, and laid the tibial graft along their forked processes. Such grafts held and immobilised the vertebræ of the diseased region. He applied the same practice to the treatment of fractures of long bones. He introduced all the devices of an expert carpenter—inlaying, wedging, dovetailing, dowelling, and tonguing—to secure a perfect co-aptation of graft to the host fragments. The graft was made to fit exactly the inlay bed prepared for it, and was bound in place by tendon. Having to serve as splints, such grafts were given considerable length, so that they might take an extensive grip of the fragments. Albee regards such inset grafts as stimulating bone production at the site of fracture. With Hey Groves, he believes live grafts are much less likely to set up infective processes than metallic plates and screws. To prevent movement and strain between the graft and host fragments, the broken limb is retained in a splint until union of the fracture is well secured.

Albee has applied bone-grafting operations to fractures and deformities of all parts of the body. He employs an inlaid bone graft to fix together the upper and lower fragments of a broken knee-cap. He seeks to correct the deformity of club-foot by transplanting a wedge taken from the cuboid on the outer side and set within an incision made into the scaphoid on the inner side. The employment of bone grafts in the treatment of fractures and of deformities is now in an experimental stage; its permanent position in surgery has still to be determined.

By far the most instructive examples of bone transplantation are those in which the fibula has been used to replace the necrosed shaft of the tibia. The particular case I am to call your attention to is the first on which Mr. C. J. Bond of Leicester operated (May 1905), and represents, therefore, the first publication of an operation of this kind.[1] In January 1903 Professor Huntingdon of the University of California had carried out the first successful substitution of fibula for tibia, publishing an account of it in 1905.[2] The same problem had presented itself to two men situated on opposite sides of the earth, and both had adopted the same method of solving it.

Mr. Bond's first case was a girl, aged 5, in whom the whole shaft of the right tibia had been destroyed by osteomyelitis eleven months previous to the date of operation. Skiagrams show that the upper tibial epiphysis, although it had persisted, had not quite escaped the effects of disease. The lower epiphysis remained—apparently healthy. In his first operation (May 1905), Mr. Bond divided the neck of the right fibula, and fixed the upper end of the shaft into the epiphysis of the tibia. The lower end of the fibula could not then be transferred to the tibial epiphysis, as there was still present in it a small area of diseased bone. Two years later, in 1907, it had so far recovered that Mr. Bond was able to separate the shaft from the external malleolus and implant it in the adjoining tibial epiphysis. The shaft of the fibula now occupied a position between the upper and lower tibial epiphyses. A skiagram taken in 1909, four years after the upper end of the fibula had been transplanted and two years after the shifting of the lower end, shows that the graft had assumed the shape and thickness of the tibia of the left or healthy leg, but was much shorter than the normal bone. That shortening is clearly due to a defect in the upper diaphyseal growth disc; although the upper end of the fibular graft has widened and formed a connection with the whole width of the old tibial epiphysis, there is no trace of the clear line which should

[1] "On the Late Results of Three Cases of Transplantation of the Fibula," *Brit. Journ. of Surg.*, 1914, vol. i. p. 610.

[2] *Ann. of Surg.*, 1905, vol. li. p. 249.

mark the site of the growth disc of the new shaft. Mr. Bond noted that the thickening—the "tibialisation"—of the new shaft began at the extremities implanted in the epiphyseal ends of the old tibia. He suspects that there was an invasion of the extremities of the new shaft by swarms of osteoblasts derived from the tibial epiphyses. However that may be, the remarkable fact remains that in the course of less than four years the slender and peculiar shaft of the fibula was, when placed between the stresses which fall on the knee and ankle joints, transformed into a tibia. At the age of 13 the girl could walk and run and "showed no noticeable limp."

In 1908 Mr. Bond performed a similar operation on a boy of 11 (see Fig. 6). The lower tibial epiphysis was so porous that the lower end of the transplanted fibula worked its way almost into the ankle-joint. Five years later, when the boy was 16 and able to walk three miles to school, a skiagram of the new tibia showed that a less successful result had been obtained than in the first case. The proximal part of the new shaft exhibited the greatest growth response, yet at its upper end it had not effected a union with more than half the width of the old tibial epiphysis. The middle part of the fibular shaft remains slender when compared with the shaft of the companion bone of the healthy leg. There can be little doubt as to the cause of this comparative failure. It lay in the condition of the lower tibial epiphysis. It was so porous that, when the fibula was implanted in it, it sunk through almost to the ankle-joint. The lower tibial epiphysis could offer neither a firm base of support, nor an osteogenetic stimulus to the fibular graft.

In 1908, soon after transplantation was effected, we can see from the skiagrams taken of this lad's leg that the lower end of the transplanted fibula had made an early effort to secure a more firm basis of support at the ankle. A tongue of ossific matter is seen to be formed in the direction of the external malleolar from which the fibula had been detached (Fig. 6). Five years later, we see from the skiagrams that the fibula has shed the lower part of the original shaft which united it to the tibial epiphysis, and acquired a new and direct prolongation to the ancient support—the external malleolus. It is when we study cases like those of Mr. Bond that we see Hunter was not so wide of the mark when he ascribed a form of "consciousness" to living bony matter.

Mr. Bond's first case also gives us another instance of "an osteogenetic adaptability." In that case—a girl of 5—the head and growth disc of the fibular shaft (upper epiphyseal

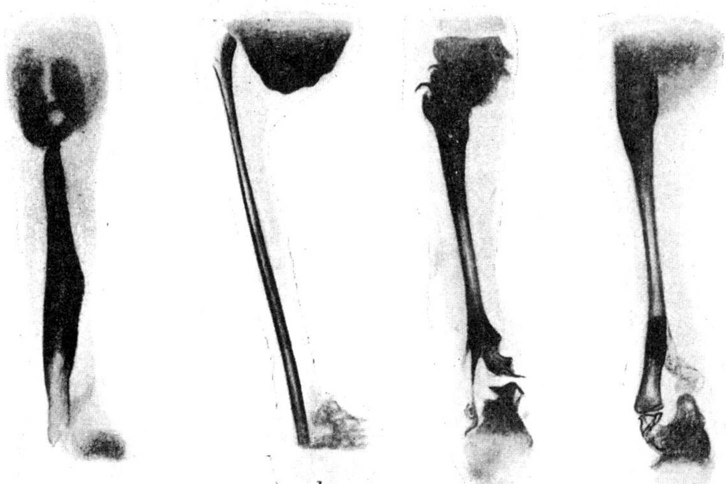

a *b* *c* *d*

FIG. 6.—Stages in the repair and growth of the transplanted fibula of Mr. Bond's case.

a. Before operation (1908). *c.* Condition in 1910.
b. After operation (1908). *d.* Condition in 1913.

line) were left attached to the outer tuberosity of the tibia. From the disc was produced a piece of shaft several inches in length. The growing distal end of the fibular head formed a union with the upper part of the new tibial shaft, and thus served as a buttress to support the upper extremity of the tibial shaft. Mr. Bond rightly supposes that there exists an attraction or "tropism" between opposed fractured bony surfaces which are undergoing repair or growth; at least one has to postulate that osteoblasts are endowed with a power of that kind. We shall inquire more closely into the endowments of osteoblasts in my next chapter.

If we sum up in the light of our present knowledge the conditions which determine the successful grafting of bones or of bone fragment, we must place first—asepsis. Grafts answer best which are taken from the patient's own body; the closer the genetic relationship of the graft-host to the recipient, the better is the graft likely to answer. The graft-host should be young, and those grafts do best which contain all three elements of bone—bone, periosteum, and medullary tissue. The graft-bed must be free from blood-clot. Washing in normal saline solution damages the vitality of a graft. A graft must be placed so that contact is effected with adjacent bony fragments; the contact must be firm and so designed that the graft becomes early subjected to the mechanical stresses and strains of the part. Anyone wishing to consult the latest original document dealing with the conditions which concern the successful use of bone graft in the repair of fractures or supplying deficiencies in the shafts of bone, should consult the conclusions come to by Mr. Hey Groves—conclusions based on an extensive series of experiments.[1]

[1] "Methods and Results of Transplantation of Bone in the Repair of Defects caused by Injury or Disease," *Brit. Journ. of Surg.*, 1917, vol. v. p. 185.

JULIUS WOLFF,
BORN 1836; DIED 1902.

CHAPTER XVIII

WOLFF'S LAW OF BONE TRANSFORMATION

In this chapter are considered the factors which regulate the shape assumed by bones under normal and abnormal conditions. Julius Wolff, who sought to formulate in words the law which regulates the morphogenesis of bones, was born in West Prussia in 1836, of a Jewish stock—at least his physical appearance leads one to draw that inference. He studied Medicine in Berlin, and taking his degree in 1860 devoted himself, at Langenbeck's suggestion, to a series of researches upon the reproduction of bone. When he commenced practice in Berlin in 1861 he applied himself to the investigation and treatment of orthopædic cases, and later in life, in 1890, when he was 54 years of age, and had established a world-wide reputation, was made Professor of Orthopædic Surgery in the University of Berlin. He died in 1902. It is possible, as we shall see, to give a much simpler statement of the law of bone growth than that given by Julius Wolff.

WE are now to examine the principles and practice of a German orthopædic surgeon, Julius Wolff of Berlin, who, in 1892, when he was 56 years of age, and had been thirty-one years in practice, published a great atlas-monograph, to which he gave the title: "The Law of Bone Transformation."[1] He succeeded in riveting the attention of thinking medical men on a mysterious property of living bone—namely, that its external form and its internal structure change with every alteration of function; in brief, a bone has the power to adapt itself to the burden it has to bear. The expression which Wolff gave to the law was new, but the substance of it takes us back to the time when Charles Bell was writing his *Illustrations of Paley's Natural Theology*, in 1834. "A bone may be taken to prove," writes Bell, "that in Nature's work strength is given with the least possible expense of materials. . . . Nature, solicitous for our safety, in a manner we could not anticipate, combines with the powerful muscular frame a dense and perfect texture of bone. . . . The inert and mechanical provision of the bone always bear relation to the living muscular power of the limb." It will be seen that the idea which underlies Wolff's law was abroad in men's minds before his time, but he was the first to give that idea a definite expression. He was the first to devote thirty years of constant work and observation to prove that the shape and structure of growing bones and adult bones depend on the stresses and strains to which they are subjected. By altering the lines of stress the shape of a bone could be changed. That, he thought, was a doctrine of hope for orthopædic surgeons.

A rapid survey of the first decade of Wolff's professional life will place us in possession of the circumstances which led him up to his discoveries—if discoveries they can be called. He was born in West Prussia in 1836, and studied Medicine in Berlin, where he graduated in 1860 at the age of 24. Amongst his teachers was Langenbeck, who served as a link between the young pupil and the surgical practices which, as we have seen in a former chapter, were initiated at Hanover by Stromeyer in 1831. In 1860, too, the problem of bone

[1] *Das Gesetz der Transformation der Knocken*, Berlin, 1892.

reproduction was much discussed; Ollier had just published his earlier researches. Langenbeck advised Wolff to make "experimental reproduction" of bone the subject of a graduation thesis. In the course of studies and experiments connected with the preparation of this thesis, Wolff became acquainted with discoveries which have been already cited in these chapters —the discoveries of Duhamel, Hunter, and Flourens. He started away in 1860 from the point at which Flourens left off in 1842. Having finished his thesis, he settled down to practice in Berlin. We can measure the progress of his knowledge from certain papers he published early in 1868—preliminary studies on the growth of bone.[1] We find there that he had broken away from the simple teaching of Flourens, who held that the shaft of a bone grew in thickness by deposition of new bone under the periosteum, accompanied by a simultaneous absorption from the walls of the medullary cavity. The periosteum, Wolff found, had but a weak osteogenetic power even in young animals, and that, when growing animals were fed on madder, newly formed bone was frequently to be found on the wall of the medullary cavity. He reverted to Duhamel's conception—that the shaft of the bone actually expanded; that by some mysterious process new bricks could be forced between the old, and the old building thus enlarged. He bored holes in the shaft of growing bone at measured distances and found, as growth proceeded, that these holes moved apart. We have seen that one of Hunter's specimens shows such a separation. He concluded that the old conception of interstitial growth in bone was true. He was constantly on the outlook for every trace he could discover of the interstitial absorption and deposition of bone, particularly in the spongy extremities of long bones, and found confirmation of his opinion. In fully grown animals, however, he believed that structural elements of a bone could, under certain circumstances, remain unchanged: so long as a bone performed exactly the same function its structural elements underwent no change.

Wolff believed that he was the first to recognise that the

[1] *Berl. Med. Wchenschr.*, 1868, vol. v. pp. 62, 76, 110.

tissue of growing bone was in a state of constant flux; for him bone was a plastic tissue in which demolition and rebuilding were taking place in all parts of a living bone. In making such a claim he was scarcely just either to Hunter or to Flourens. We have seen that Hunter's conception of bone growth was largely based on a study of the process as seen in the upper extremity of the femur—particularly of the head and neck. Hunter discovered that the elaborate architecture of the neck was being continually remodelled from infancy to adult life; without such remodelling, the femoral shaft could not grow. The following passage shows how changeable a substance he conceived bone to be: "Bones begin at a point and shoot out at their surface, and the part that seems already formed is not in reality so, for it is forming every day by having new matter thrown into it, till the whole substance is complete; even then it is constantly changing its matter."[1] The following passage will show that Flourens held a similar conception of living bone: "Mais si, d'une part, des molécules nouvelles sont incessamment déposées, si, d'autre part, des molécules anciennes sont incessamment résorbés; il y a donc mutation continuelle de la matière. La mutation continuelle de la matière est la résultat général et le résultat le plus important de toutes les expériences de cet ouvrage."[2] Wolff's realisation that living bone was a plastic tissue was not new; but we shall see he applied this knowledge in a way and with an insistence no one had ever done before.

A study of the neck of the femur, which proved so fruitful to Hunter, gave Wolff the key to his law of transformation. It is strange that the elaborate architecture of the neck of the human femur failed to attract the attention of early anatomists. Almost the first description we have of its architecture—the very first attempt to probe its mechanism—is that given by a contemporary of John Hilton, Mr. F. O. Ward, who demonstrated anatomy at King's College, London. The concise treatise on *Human Osteology*, which was first published in 1838, and found to be a very dull book by many

[1] *Collected Works*, vol. ii. p. 18.

[2] *Recherches sur le Développement des Os*, 1842, p. 98.

of those who came of a later generation, is packed with first-hand observation. Many of us will remember the drawing of a triangular bracket supporting a street lamp, which was inserted by Mr. Ward to explain the architecture of the femoral neck. The bracket, which supported the lamp from an upright, was made of two pieces—an oblique supporting and a cross tie-piece. The architecture of the femoral neck showed two corresponding sets of lines or trabeculæ—an ascending set springing from the lower wall of the neck to the head, and a horizontal or cross set occupying the upper part of the neck. Ward supposed that when a man stood up these two sets of trabeculæ transferred the weight of the trunk to the shafts of the thigh-bones; the ascending lines were then subjected to pressure, the horizontal lines to tension.

In 1867, when Wolff was busy with his investigations, and such practice as then came his way, the architecture of the femur was being subjected to a thorough analysis by two professors in Zurich—Von Meyer, the anatomist, and Culmann, engineer and mathematician. Culmann had an intimate acquaintance, both practical and theoretical, with the problems which engineers have to solve in designing iron bridges and cranes. On examining Von Meyer's preparation and drawings of the femoral neck, he was convinced that, in a mathematical and engineering sense, the architecture of the femoral neck was a perfect mathematical design for the transference of weight. The trabeculæ had been given the exact position and form which the expert mathematician and engineer would postulate for them. Culmann saw in the architecture of the upper extremity of the femur the counterpart of the design which Fairbairn had embodied in his crane. Modern examination[1] of the structure of the femoral neck shows how exact Culmann's comparison was. Here, then, we have a very remarkable discovery: the osteoblasts are the architects of the femoral neck; they are engineers who do their work according to "exact mathematical laws."

[1] See Professor Francis Dixon's paper, *Journ. of Anat. and Physiol.*, 1910, vol. xliv. p. 223.

Having absorbed the Zurich discovery, we see that Wolff began to extend his knowledge in several directions. He commenced to formulate his law of bone transformation in 1870, but kept adding item upon item to it until 1885, when he satisfied himself it was complete. The first postulate which he embodied in his law was founded on a study of sections of deformed bones; he observed that when a bone became deformed its internal structure was radically changed. The internal changes, he concluded, were due to an alteration of the static function, an alteration which must take place in every deformed bone; a deformed bone ultimately became perfectly adapted to its new function or position, "according to mathematical law." He further observed that these primary alterations in the internal structure of a bone were followed by certain secondary adaptative changes in the external form, "according to mathematical law." Then in 1884–1885 were added the two final and more important clauses of his law, namely: If a *normal* bone is used in a new way its structure and form will change to meet its new function; if a deformed bone is rectified, and its normal function thus restored, then that bone will reassume and retain its normal shape and structure. In its final form Wolff's law of bone transformation came to read thus—I give the usually adopted translation—"Every change in the form and the function of a bone or of their function alone, is followed by certain definite changes in their internal architecture, and equally definite secondary alterations in their external conformation, in accordance with mathematical laws."

In framing the definition of his law of bone transformation Wolff buried in words a simple and vital truth. The late Dr. John B. Murphy of Chicago found a much simpler expression when he said: "The amount of growth in a bone depends upon the need for it." To make quite certain we shall not miss the more important truths which Wolff embodied in his law, we shall apply it to a concrete instance—knock-knee or genu valgum. We want to know the cause of genu valgum, so that we may know how to prevent the occurrence of the deformity. Wolff held that, in the majority of cases, the deformity was due to keeping the lower limb in a faulty posture; he cites as an instance the apprentice baker standing over the dough trough with legs apart. Wolff regarded the living bone as a sensitive plastic tissue, controlled by the lines of force transmitted by the framework of the limb; as long as the lines of force passed in the normal axis of the limb the bones retained their normal form and architecture. But if the lines of pressure are deflected, as in the straddled limbs of the apprentice baker, then the living bones will begin to alter the disposition of their particles, rearranging them to meet the new lines of transmission. It is rearrangement of the living particles of the bone which produces knock-knee. It was Wolff's belief that, save in those few cases where there is softening of the bones or actual disease of the limbs, no one would become the subject of knock-knee if he or she stood aright. Knock-knee was, in his opinion, the result of a faulty posture.

There can be no doubt of the extensive structural changes which take place in the bones of the lower extremity in case of knock-knee of long standing; Wolff has given the most complete proof of that and also of the fact that the alterations are just such as will meet the altered lines of pressure. There are few, however, who will agree that the cause of knock-knee is so simple as Wolff here supposes. In most cases we fail to find a history of faulty posture, but we do discover in the majority of such cases that there have been prolonged periods of standing or of standing and walking. We are driven, as I have pointed out in a previous chapter,[1] to look for the primary cause, not in the bones, but in the muscles, particularly in those which are tonically and constantly in action so long as we are standing. It is when such muscles give way that the bony changes, which have been depicted by Wolff, set in. Wolff rejected with scorn any attempt to lay the primary blame of deformities on muscles. And yet the evidence all points to muscles as being the primary defaulters; in all static deformities I believe that the transformation of bone is a direct result of defective or

[1] Chapter X.

unbalanced muscular action. The principles which guided Wolff in the treatment of deformities were based on his law—on the response which living bone makes when submitted to pressures. In the case of knock-knee, his first step in treatment was to restore the normal lines of pressure to the tibia. That can be done only by bringing the tibia back into its proper alignment with the femur—either by performing Macewen's osteotomy of the femur, or, as Wolff preferred, by exerting pressure on the region of the knee through the application of splints and bandages. These were forms of treatment in use long before orthopædic surgeons had heard of the law of bone transformation; the law did not lead to any change in our remedial methods. Wolff, however, was the first to bring home to surgeons and anatomists the extensive nature of the changes which had to take place in the internal architecture of the femur, tibia, and fibula, after rectification for knock-knee, or any deformity, was effected. The internal structure, which had been changed to suit the transmission of force in an abnormal direction, had to be remodelled to meet the demands of a normal transmission.

We are now to take our leave of Wolff. He had an uphill struggle in Berlin; recognition came late to him. In 1890, when 54 years of age, he was appointed Professor of Orthopædic Surgery in the University of Berlin. He died in February 1902 at the age of 66.

Every student who has read Wolff's monograph with care must have noted that at no time was he concerned with the actual bone-builders—with the osteoblasts themselves. He has given us excellent plates illustrating the marvellous manner in which the internal structure of deformed bones has been remodelled to meet new lines of pressure; but nowhere does he mention the cunning engineers. His monograph, and the same may truly be said of his law, is a stage set out with all the necessary fittings for a play, but the actors are never called on to appear. We are now to turn to these actors—the osteoblasts—and see if we can obtain a closer knowledge of the remarkable engineering powers with which they are apparently endowed.

In a former chapter we have mentioned Goodsir's investigation of a sponge from Spitzbergen, and the discovery of the cells or scleroblasts which formed its skeleton of spicules. We shall see when we come to deal with these spicule-builders that we have before us ancient, ancestral relatives of osteoblasts, so ancient that we have to go back to the deeper strata of the earth to find their first appearance in the animal phylum. Goodsir's investigations were made in 1852; at the end of the century the late lamente.l Professor Minchin, when he held the chair of Zoology at University College, London, again took up the study of scleroblasts. The investigations, however, with which I shall now deal are those made by a pupil of his, Dr. W. Woodland.[1] In 1904 Dr. Woodland began the study of spicule-building in one of our commoner sponges (*Sycon coronata*), which grows in shallow tidal waters or is planted on rocks and exposed to the action of falling waves. It is important for us to note that this sponge, studded with calcareous spicules, is of a cylindrical shape, tapering to its attached base and also to its free or oscular extremity. There is a central or gastral cavity surrounded by the sponge wall. The sponge wall, being made up of soft tissue—an outer layer of dermal cells and an inner of lining or gastral cells—requires support and protection, which are supplied by a skeleton of separate calcareous spicules. The soft walls of this sponge are constantly exposed to the force of moving waters, and we shall see that the spicule-builders—the scleroblasts—are endowed with the same properties as osteoblasts—the powers of fashioning and depositing the elements of the skeleton so that the sponge can best resist the forces to which it is habitually exposed.

We shall first follow the manufacture of the simpler form of spicule—the needle-shaped or monaxon type. Dr. Woodland found that the scleroblast which was to give rise to such a spicule moved inwards from the covering or dermal layer of cells and took up a position in the sponge wall. The scleroblast divides, and at the beginning of the process a calcareous spicule commenced to form—to be secreted—

[1] "Studies in Spicule Formation," *Quart. Journ. Micr. Sc.*, London, 1906, vol. xlix. pp. 231, 533.

within the cell substance. Presently, as the spicule grows in length the daughter scleroblasts separate, each moving towards opposite ends of the spicule. The scleroblasts are now applied to only one side of the spicule, which has kept growing in length until its outer end projects into the dermal layer. Then the outer scleroblast forsakes the spicule and moves within the sponge wall, and shortly afterwards, accompanied by its companion, disappears. The needle-shaped spicules thus formed may project, and are often shed, from the wall of the sponge, and are evidently intended for protection as much as for support.

Besides these simple needle-shaped spicules the same sponge is also provided with others of a more complex type—triradiate and quadriradiate—which are deftly planted in the sponge wall so as to give it a maximum strength. Dr. Woodland found that these more complex spicules were formed by groups of scleroblasts which arranged themselves—as Goodsir had observed—in definite groupings. Their combined efforts produced the same perfectly fashioned piece of masonry every time. Dr. Woodland observed that if the position of a sponge was altered, so that the side exposed to the force of waves was reversed, the scleroblasts reacted in such a way as to strengthen the side on which the chief stresses now fell. Here then is Wolff's law regulating the activities of the skeleton-builders of sponges. We must suppose that such cells are sensitive to certain forces—the pressures and tensions caused by moving waters and also to the force of gravity.

Lately, Professor Dendy, of King's College, London, working in partnership with his mathematical colleague, Professor Nicholson, has discovered a curious fact relating to the behaviour of spicule-forming scleroblasts. The sponge investigated by Professor Dendy is provided with long fusiform spicules. The scleroblasts, having completed the growth of a spicule as regards length, become attached at certain intermediate points of the spicular shaft, where they deposit an encircling whorl. Professor Nicholson has shown that the points at which the scleroblasts attach themselves to form whorls correspond to the internodes, the non-vibrating points, of such spicules. Apparently, as Professor Dendy has supposed, the scleroblasts are so sensitive to vibration that they are able to discover the points of least vibration on any given spicule.

In the various sponges which we have selected here as examples, the scleroblasts work their spicules into the sponge wall, much in the same manner as a corset-maker inserts supports of steel or whalebone into stays. In other sponges, however, they work in the manner of osteoblasts; the product of one group of builders is united with that of a neighbouring group, so that a framework or skeleton is laid down "according to mathematical laws" and designed to meet the stresses to which their building is exposed. There is no more beautiful engineering feat than that to be seen in the Zeppelin-like framework of the Japanese sponge, Euplectella. No one who has watched the behaviour of scleroblasts and marked the design in their workmanship can doubt they have acquired certain characteristic qualities—chief of which is a sensitiveness to vibrations—to stresses. We see them build the same form of spicules as their ancestors, and therefore must suppose that the building quality is a gift handed on by inheritance. We see them alter their mode of building as stresses change; we must, therefore, suppose that their inherited powers can be changed by the circumstances under which they work.

The importance of these observations on spicule-building on our interpretation of the behaviour of osteoblasts—on our interpretation of Wolff's law—must be apparent to every orthopædic surgeon. Osteoblasts seem to conduct the work of bone-building as if they had been given the training of expert and unerring engineers: as a child grows they keep rebuilding the articular extremities of bones; the architecture of the neck of the femur is being constantly remodelled, but the osteoblasts maintain the same complex design throughout. When an angular union results from fracture, they rearrange the substance of the bones in accordance with the new lines of pressure, removing the parts which are no longer subjected to direct stress, and laying down new supports in the lines where pressure is active. If a bone becomes soft and bends, they meet the new condition by rebuilding along its cavity. When we speak of Wolff's law we really mean the law of osteoblasts; it would be a great and direct gain to practical surgeons if they would discard the clumsy expression given to this law by its author. The law is simply this: Osteoblasts at all times build and unbuild according to the stresses to which they are subjected.

It is not Wolff's law which an orthopædic surgeon should cultivate, but a close acquaintance with the behaviour of osteoblasts. If he understands them fully and treats them rightly, they can prove his most helpful friends; if he ignores them, they can and do prove obstreperous. In the first place, as we have seen from their behaviour in grafts, they must be subjected to stresses; stress is as necessary for their health and activity as exercise is for the living body. The mode in which they build and the lines along which they will deposit their building material are determined by the forces which are brought to bear on them. They can, by manipulating such force, be made to build as we will. We see further that they can be stimulated to overwork in many ways. H. O. Thomas, by beating the external condyle of the femur with a mallet, made the bone cells build up a rampart which prevented the outward dislocation of the patella. Injury to a bone excites every osteoblast in that bone; diaphyseal growth discs act as buffers in limiting the radiation of a stimulus; acute inflammation increases their activity and destroys their sensitiveness to pressures.

The study of diseased conditions brings out other qualities of osteoblasts. In giants they become stimulated to an extraordinary degree by a certain substance or substances derived from the pituitary body; in certain forms of dwarfs we see this elixir of growth withheld. In acromegaly we see growth again reawakened under a pituitary derivative—a growth which chiefly affects cartilage-covered areas of bone—areas which are constantly subjected to movement. In Paget's disease we see quite another disturbance in the life of osteoblasts. They are no longer acutely sensitive to the stresses which fall on the skeleton; they lose their engineering qualities and lay down their materials clumsily.

The study of many other conditions—such as achondoplasia, periosteal dystrophy, rickets, mollities ossium, fragilitas ossium—throw side-lights on the complex nature of osteoblasts.

Wolff did orthopædic surgery a real service by insisting on and establishing the plasticity of bone, but his law is a clumsy expression of but one of the vital properties of osteoblasts. Further, we see that Hunter was not very far off the truth when he attributed a form of "consciousness" to living bone.

PROFESSOR PETER REDFERN, 1821–1912,
Founder of our knowledge concerning the microscopic structure of cartilage and discoverer of the process by which its wounds are repaired.

CHAPTER XIX

THE ORIGIN OF OUR KNOWLEDGE CONCERNING ARTICULAR CARTILAGE IN HEALTH AND DISEASE

In this chapter the origin of our knowledge concerning the growth and repair of articular cartilage is traced. Joseph Toynbee, born 1815 in Lincolnshire, died 1866 in London at the age of 51, is best known for his pioneer work in aural surgery, but his first investigation, made in 1838, which is dealt with in this chapter, was concerned with the manner in which articular cartilage is nourished. Dr. Peter Redfern published the most important account of his researches on cartilage in 1850 while Professor of Anatomy and Physiology in King's College, Aberdeen. He was born in 1821 at Chesterfield, Derbyshire ; studied at Edinburgh, where he heard Goodsir lecture on cartilage in 1842. After taking his medical degree at London University in 1844, he was called to Aberdeen, where he remained until 1860, when he became Professor of Anatomy and Physiology in Queen's College, Belfast. He retired in 1893, and died at Donaghadee, Ireland, in

1912, aged 91. Sir Alexander Ogston, emeritus Professor of Surgery in the University of Aberdeen, was born in that city in 1844, the son of Dr. Francis Ogston, Professor of Medical Jurisprudence.

HITHERTO we have been dealing with osteoblasts; we shall now consider, from a surgeon's point of view, cells of a totally different kind—cartilage cells or chondroblasts. A surgeon cannot make a greater mistake than to suppose that these two kinds of cells are closely akin: there is the same relationship between them as between the wolf and the lamb; the one feeds on the other whenever they are set in unrestrained proximity.

We have only to note how differently bone cells and cartilage cells react to injury to realise how opposite they are in nature. When the femur is broken the osteoblasts at the broken ends seek each other out and repair the breach; so strong is their propensity to effect a union that only the most unfavourable circumstances can frustrate their instinctive efforts. When a costal cartilage is fractured or cut we see a totally different result; cartilage cells have no tendency to grow, to seek each other out, and to repair a breach. Their innate tendency is of quite a different kind; when irritated, they divide, enlarge, and then disappear by melting. For its repair, cartilage has to depend on outside help; fibroblasts have to be called in to repair the breach.

When we note the exact manner in which a joint, such as that of the knee, appears in the budding limb of the embryo, we may, without any stretch of imagination, describe its formation as a breach or fracture produced by developmental means. It is produced by a cleft or rift appearing in the skeletal bar; the femoral roof of the cleft, its tibial floor, its capsular sides are lined with cartilage. In time centres for ossification appear in the adjacent epiphyses of the femur and tibia; these centres invade the cartilage and progress towards the joint, but never reach it, because the growth of the articular cartilages keeps an even pace with the progress of ossification. Finally comes adult life, when the osteoblastic invasion of cartilage, by some process which is still obscure, is brought to a standstill. We have, then, between the adjacent bony surfaces of the femur and tibia containing their armies of osteoblasts, two strata of articular cartilage which collectively form a barrier some two-fifths of an inch (10 mm.) in thickness. The opposed femoral and tibial cartilage cells have no inclination or power to unite; but that is not the case with the corresponding armies of osteoblasts lying behind the cartilage. Articular cartilage, so long as it is healthy, forms a barrier which osteoblasts cannot penetrate. But if the cartilaginous barriers be destroyed by infection entering the joint-cavity, or the adjacent epiphyses, or by an obscure nutritional or rheumatoid change, then a joint becomes reduced to the condition of a fracture. Healing may result in the production of a "false-joint," a fibrous union or a true bony union. The reason why bony union is so difficult to obtain is best explained by an observation made by Macewen. He noted that the osteoblasts of epiphyses have a lower osteogenetic power than those of the shaft. However that may be, there can be no doubt that a knowledge of articular cartilage —a knowledge of chondroblasts—is the only rational basis for joint therapy.

By way of introduction we have been drawing attention to the radical differences which exist between the cells which build bones and those which compose the chief substance of joints. We are now to set out on one of our usual journeys in quest of some of the men who have helped to lay the modern foundations of our knowledge of joints. For the first, we need not go beyond the walls of the Royal College of Surgeons. In 1838, just as the cell doctrine was being incubated in Berlin, Joseph Toynbee, duly qualified as a medical practitioner, and 23 years of age, came to assist Clift and Owen in the work of the College Museum. Toynbee was one of the first in this country to apply the microscope to unravelling the cellular constitution of living tissues. His attention was particularly directed to non-vascular tissues—tissues which lie outside the blood circulation, such as the cornea, the lens, the vitreous humour, dentine, articular cartilage, and the epidermis.[1] He

[1] "The Non-vascularity of Certain Animal Tissues," *Phil. Trans.*, 1841, Part II. p. 159.

discovered the cellular nature of the epidermis, but it is his pioneer work on the nourishment of articular cartilage which concerns us here. Immediately under the thin lamina of bone, on which articular cartilage is implanted at the ends of long bones, he found a rich carpet of looped vessels, each loop being provided with one or more clearly marked dilatation. He inferred that this vascular carpet was a special provision for the nourishment of the articular cartilage. The cartilage cells imbibed their nourishment from the lymph thrown out by the sub-articular plexus; the lymph, he supposed, soaked into the cartilage, for nowhere could he detect open pores or channels through which it could flow. He also noted, under the synovial membrane, at the margin of the cartilage, the remarkable vascular plexus first described by Wm. Hunter in 1743. The peri-articular vascular circle also aided in furnishing lymph for the cartilage. We are apt to think of Joseph Toynbee as the founder of aural surgery and forget that he also established one of the fundamental facts relating to joints —namely, that their cartilages lie outside the circulation and are nourished by exuded lymph. Toynbee's observation gives the modern operating surgeon an assurance that articular cartilage ought to be a substance eminently capable of being transplanted. In its natural state it may be said to form insets or grafts rather than intrinsic parts of the animal body.

Toynbee's research was published in 1841. To find the next important addition to our knowledge of cartilage we must visit Edinburgh in 1842, when young Goodsir became curator of the Museum of the College of Surgeons. We have already noted how quickly he grasped the real nature of bone; his genius is equally apparent when he turned to the investigation of cartilage. Hunter had taught that absorption of tissues was carried out by lymphatic vessels. Goodsir could find no lymphatics in articular cartilage, and yet, as he knew, it could undergo absorption. Clearly, the lymphatics could not be the only elements concerned in absorption. To settle the matter, he chose for investigation a joint which suffered from caries (tuberculosis). Cutting microscopic sections from the articular cartilage at the junction of the diseased and healthy parts, he found that the pits in the cartilage—the areas undergoing absorption—were filled with a cellulo-vascular tissue. He satisfied himself that these invading cells were demolishing the cartilage. He observed that, as the invaders approached, the cartilage cells multiplied, became large, rounded, and swollen just as they did in the cartilage of the epiphyseal line when the osteoblasts and their vessels invaded their territory. He saw that such enlarged cartilage cells ultimately ruptured, and that the invading cells occupied the spaces thus made. He believed that the cartilage cells actually joined the invaders. Goodsir traced the origin of the "nucleated vascular membrane" which thus invaded and demolished articular cartilage to the vascular margin—the peri-articular plexus of the joint. The vascular membrane was an adventitious or invasive tissue. He also observed that, in the disease of joints, articular cartilage might be attacked from its deep or osseous surface, as well as on its free surface. Eruptions might break through from Toynbee's sub-articular plexus, and the deep and superficial invasions might meet and fuse within the substance of the cartilage. By such a process the whole area of a cartilaginous articular surface might be destroyed and replaced by a fibro-vascular tissue.

If the process was plain to Goodsir, its significance certainly remained a mystery. The secrets of suppuration and infection were unknown in his time. Yet anyone who has watched a pannus spread across the cornea to reach an injured or infected point has before him all the outward appearance which Goodsir recognised in tubercular joints. A non-vascular tissue like articular cartilage is defenceless in the face of an invasion of micro-organisms. On its deep surface it is protected, for its basal layer forms so fine a filter that we cannot conceive germs entering it from that side. Its free surface, suffused with a nutritive medium—the synovial fluid—offers, on the other hand, a wide and favourable field for such pathogenic germs as may reach the interior of the joint. Under such circumstances the articular cartilage depends on the vascular tissue at its margin for defence.

The vascular tissue responds by invasion; Goodsir observed the price that articular cartilage has to pay for its non-vascularless, defenceless condition. The explanation of that condition we shall see later.

We have now to pass farther north, to the University of Aberdeen, to see the next important addition made to our fundamental knowledge of articular cartilage. The introduction of the compound microscope gave the young men of the "forties" of last century their opportunity. Toynbee was young then, so was Goodsir, and the man we are now to meet —Peter Redfern—was still youthful when he did his pioneer work on cartilage.[1] He was a Derbyshire boy, born in 1821, and began his medical education in Edinburgh, completing it at University College, London, in 1844. In the following year, when 23 years of age, he was called to King's College, Aberdeen, to fill the chair of Anatomy and Physiology. It was some years after his arrival in that northern city that he began the researches which concern us. He obtained material from the Infirmary and made an accurate microscopic survey of the articular cartilage from thirteen patients who suffered from disease of their joints—four of them being cases of chronic rheumatism. He followed up these investigations by a series of ninety experiments on the healing of cartilage—performed chiefly upon dogs. These experiments began in 1849 and were finished in 1851, and show an uncommon ability for research. One marvels how it came about that Professor Redfern, who was destined to outlive all the pioneers who made the microscope a power in Medicine—he died in 1913 at the age of 91—left the research which we are now merely to glance at as almost his sole contribution to knowledge.

We are chiefly interested in the observations he gathered from his experiments on animals. Incisions which he made in the articular cartilage of the patella or of the condyles of the femur, remained open and unhealed for many months,

[1] *A Normal Nutrition in Human Articular Cartilage*, Edinburgh, 1850. "On the Healing of Wounds in Articular Cartilage," *Month. Journ. Med. Sc.*, 1851, vol. xiii. p. 201.

and yet the animal could use the joint normally and never manifested any sign of pain. When an incision in the cartilage did ultimately heal, repair was effected by means of fibrous tissue, not by the reproduction of cartilage. Redfern believed that the fibroblasts which filled the breach were derived from neighbouring cells — were modified cartilage cells; but we note that his incisions which reached or nearly reached the margin of the cartilage healed much sooner than those which were isolated and placed at some distance from the peri-articular tissue. Repair, we suspect, was effected by an invasion from the vascular fringe. He noted that only the cells in the immediate vicinity of the incision or on the edges of the breach made by applying a cautery showed any reaction—multiplication and enlargement. Cells lying 3 mm. beyond the area of destruction showed no reaction. In this respect articular cartilage differs from all other tissues. In vascularised tissues the reaction extends far beyond the site of injury, but articular cartilage is a loosely organised community of separated cells; neighbours make little or no response when their fellows are injured. If he removed part of a dog's limb by amputating at a joint he observed that the cartilage covering the articular end of the stump became pitted beneath the flaps which covered it. From Redfern we learn that articular cartilage, unlike bone, has no power to effect repair by reproduction of its own substance.

The experiments which Redfern carried out on costal cartilage are particularly instructive. One of these was to thrust a needle threaded with a silk ligature through a cartilage and then tie the ligature so as to enclose half the thickness of the cartilage within its grip. In three days the cartilage cells within the ligature had died; the ligature came away spontaneously, and an open fissure was thus produced. The fissure thus occasioned always deepened until only the perichondrium on the pleural aspect held the ends together. The gap was filled up in time by fibrous tissue, never by cartilage. If he included the entire thickness of the cartilage within his ligature, all within the ligature disappeared in four days, a gap remaining in the cartilage which was never filled.

From this series of experiments we can draw three important inferences: (1) that if we apply continuous pressure to cartilage cells they will die in three days; (2) that in the process of dying the cells and their intercellular tissue melt and disappear; (3) that cartilage cells are destitute of the powers of repair.

His observations on the microscopic structure of cartilage affected by the condition which passes to-day under the name of chronic rheumatic arthritis are the earliest we have and cannot be passed by without mention. He was familiar with the three strata of cells which can be recognised in every section of articular cartilage. The deepest stratum, next the bone, has its cells arranged in vertical columns (Fig. 7); the superficial stratum has its cells in horizontal lines which lie parallel to the articular surface. The middle stratum is provided with large cells, which are arranged in irregular groups. In rheumatic joints Redfern found (Fig. 7) that the superficial stratum or zone was greatly thickened and its structure completely changed. It had become a rugged fibro-cartilage. The cells of the middle zone or stratum had formed large groups of swollen cells. The interpretation of these appearances we shall try to explain presently. Meantime, we note that Redfern attributed them to a disturbed nutrition of the cartilage.

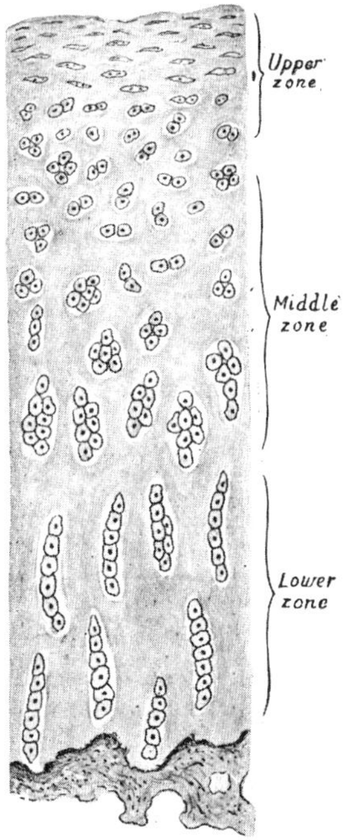

FIG. 7.—Vertical section of articular cartilage to show the three strata of cells—adapted from illustrations given by Redfern and by Ogston.

It was by a mere coincidence that the next important addition to our knowledge of articular cartilage in this country was made in Aberdeen. Long before Sir Alexander Ogston became a medical student, Profesor Redfern had vacated the chair in Aberdeen to occupy one in Belfast. We have already seen that Ogston introduced an ingenious operation for knock-knee in 1875, and it was while planning this operation that his attention was drawn to an essential point in the physiology of joints. The question he put to himself was this: How is the wear and tear of articular cartilage made good? All mechanical joints wear out by friction; why do the joints of the animal body not wear out by use? One would have thought the question which young Ogston put to himself in 1875 was one which had been answered long before that date. An account of the researches on which his answer is based will be found in the *Journal of Anatomy and Physiology*.[1] Ogston, as his predecessors had done, noted the three zones of articular cartilage—the basal with its vertical columns of cells reaching down to the bone, the superficial with its cells in horizontal strata, and the middle with its irregular groups of cells. Ogston interpreted these appearances to mean that the middle zone was germinal or reproductive in nature; it was a growth zone which fed both the superficial and the deep zone. The additions to the surface zone made good the frictional waste which resulted from movement. He conceived that the deep zone also required repletion, because, all through life, the deeper cells of the column were being taken up by the underlying stratum of epiphyseal bone. He conceived that growth, rearrangement, and absorption were processes which went on in the epiphyseal ends of bones all through life. He thought that the appearances seen in diseased joints, particularly those affected with rheumatoid arthritis, could not be accounted for unless it was presumed that the basal layer of articular cartilage was being continually, if slowly, consumed by the bone.

At the present time the problem of repair of joints is much in the position it was left in by Ogston forty years ago. It is closely linked with another important problem: "How are our joints lubricated?" Engineers have striven hard to invent a perfect system of lubrication. The ideal system of which they are in search is one to surround the revolving axle by a fine and uniform film of lubricant oil. Pressure, however,

[1] *Journ. Anat. and Physiol.*, 1876, vol. x. p. 49; 1878, vol. xii. p. 503.

causes the axle to work against one or other side of the bush, thus squeezing out the oil from the area at which the chief friction takes place. The difficulty has been partially overcome by forcing the lubricant into a joint at a sufficient pressure to separate the axle from the bush by a uniform film on all sides. But even with an ideal lubricating system metallic joints wear out. In this case, as in so many others, Nature has successfully solved the problem and secured the ideal system; our joints are furnished with cartilaginous bushes which, in the absence of disease, never wear out and which at all times are coated with a uniform film of perfect lubricant.

We must first of all form a definite conception of how growth takes place in cartilage if we would rightly understand how the waste in joints from friction is made good. The examination of a section across a costal cartilage provides us with a clue. The cells in the centre or core of the cartilage are large and swollen; a cartilage cell in that condition is one about to undergo disruption—liquefaction. We can give a rational interpretation of the appearances presented by such a section only if we suppose that the zone of cartilage cells which lies immediately under the perichondrium serves as a cambium or growth zone from which fresh broods of cells are slowly produced. From this circumferential growth zone cells pass in two directions—from its outer or superficial aspect cells are added to the perichondrium; from its inner or deep aspect cells are added to the central area or core of the cartilage. These two broods of cells—the "central" and "peripheral"—undergo different fates: the central multiply, enlarge, and deliquesce, thus providing a highly elastic core for the cartilages in youth; the peripheral, as they become embedded in the strata of the perichondrium, shrivel, become mere streaks of granules, and finally disappear. We have seen from Redfern's experiments that the natural termination of cartilage cells is deliquescence.

Growth and multiplication of cells is an extremely slow process in cartilage—slower than in any other tissue of the body; we must also presume that absorption, as in bone, is also at work, the one process balancing the other in cartilaginous structures of the adult. If we now apply to articular cartilage the interpretation which has been suggested by the appearance of costal cartilage we arrive at the following conclusions:

A vertical section of articular cartilage reveals, as we have already seen, three zones—a basal with vertical columns of cells, a superficial with horizontal strata, and a middle with large cells in irregular groups. Such an arrangement of cells suggests that cartilage cells, like bone cells, are sensitive to the forces which act through the axial skeleton of the limbs. Those of the superficial zone are subject to the frictional or rubbing effects developed at the articular surface and are arranged horizontally; the basal zone, with its vertical column of cells, is concerned in the transmission of weight from the cartilage to the bone; while the middle zone, with its irregular groups of cells, is subjected to incident stresses of a mixed kind.

When the cells of the superficial zone are closely examined it is seen that they present the same form and the same graduated arrangement as those of the perichondrium. The surface of healthy articular cartilage is not really smooth and polished; under the microscope its free margin presents an uneven or serrated outline at which a process of erosion, or, to speak more correctly, liquefaction is taking place. The surface film of the cartilage, made up of the granular débris of cartilage cells and of intracellular substance, is being worn away—rubbed down so as to form a joint lubricant. The cells of the superficial zone, then, are being continually worn away, as Ogston supposed, and by their liquefaction are providing a perfect film of lubricants over the whole articular surface.

When we examine the form and the arrangement of the cells of the middle zone we cannot fail to be impressed by the resemblances they present to those of the core zone of a costal cartilage. The middle zone of an articular cartilage is one in which the cells divide, enlarge, become swollen, and disappear. The remarkable resilience of articular cartilage is due to the presence of the intermediate zone. It is a soft elastic cushion inserted between the superficial and basal zones. It is a zone, not of production as Ogston postulated, but one of deliquescence. Its cells are apparently derived from the deepest stratum of the superficial zone which we therefore regard as a cambial or growth layer. On one side it makes good the waste in the surface stratum due to friction; on the other, the absorption or liquefaction which takes place in the middle zone. It is also possible that cells are added to the middle zone from the ends of the vertical columns of the basal zone.

We have been discussing the manner in which the frictional waste of joints is made good; the degree of waste will depend on the amount of work done by the joints. We now turn to the system of lubrication, selecting the knee-joint as a basis for our discussion. Only small areas of the opposed articular surfaces of the femur and tibia are in actual contact at any stage of a movement; the areas not in contact are covered by the synovial membrane. A synovial membrane is a cartilage covered and therefore a lubricating structure. Every area of the articular surfaces, both before and after it has been in frictional contact, comes against the synovial lining membrane of the joint. In order to make certain that the synovial membrane will fill the various crevices and inequalities, which alter at each phase of the moving knee, it has been set on soft pads of semi-fluid fat and thrown into flexible folds and fringes. The hip-joint is furnished with a "Haversian bursa" of fat, and the knee is provided with a subpatellar pad. A vacuum cannot occur in these or in any other joint, because synovial-covered semi-fluid masses of fat pass in and fill every vacant cranny. The surface of the synovial membrane is plastered over by a delicate layer of cartilage of a cellular texture, recalling in character the appearance assumed by a fragment of costal cartilage which has been grafted for some time in subcutaneous tissue. No one doubts that synovia, the lubricant fluid of joints, is formed by a liquefaction of the liquid cells of the synovial membrane. They are the chief source of the lubricant, which is applied to articular surfaces as they rub against the membrane. But I do not think that the membrane is the only source of supply. The end product of the superficial cells of the articular cartilage is also synovia, produced in the form of a lubricant surface film. Every movement of the joint produces a new film as it rubs off the old. How far synovial fluid may serve to nourish the cartilage of the joint we do not know. We know that when it is infected it becomes an excellent medium for the cultivation of micro-organisms. Normally it is absorbed by the vascular fringes of the lining membrane of the joint.

When we apply the conception of growth and lubrication, just outlined, to the microscopic appearance presented by the articular cartilage in cases of chronic rheumatic arthritis we find that many of the pathological changes receive for the first time a rational explanation. In all chronic forms of arthritis, a condition of the articular cartilage and of the synovial membrane is produced which is exactly similar to that disease of the skin we name ichthyosis: the superficial epithelium adheres and the epidermis thickens because their cells are no longer eliminated by cornification and shedding. In arthritis the superficial stratum of the joint surface no longer ripens into synovia; the nutritional system of the joint—a system we are ignorant of—is so far damaged, or the nutritional fluid supplied from the blood so altered, that the articular zone becomes fibrous and cannot be rubbed off and liquefied by normal frictional movements. The middle zone of the articular cartilage becomes studded with overgrown broods of cells. Presently the frictional forces generated between the fibrous and roughened surfaces of the articular cartilages of the joint alter the stresses which normally fall on the adjacent osteoblasts of the epiphyses. We have seen how sensitive osteoblasts are to pathological or abnormal stress. They react and produce the irregularities of the articular ends of bones—the irregularities so well described by Ogston and Arbuthnot Lane.

We have left but little space to bring our knowledge of articular cartilage up to date. Its experimental transplantation was first investigated by Ollier,[1] and almost at the same time by that experimental genius Paul Bert.[2] Ollier found that grafts of perichondrium produced no cartilage, nor would grafts destitute of perichondrium live when transplanted from one animal to another. Cartilage provided with its perichon-

[1] *Traité Expérimental et Clinique de la Régénération des Os*, Paris, 1868.

[2] *Recherches Expérimentales de la Vitalité propre des Tissus Animaux*, Paris, 1866.

drium, he found, could be successfully grafted, but the graft never reproduced cartilage. He found that slices of articular cartilage underwent a fatty degeneration when grafted in subcutaneous tissue. Cartilaginous growth discs from the limb bones of young animals never proceeded to reproduce bone when he grafted them. Bert transplanted spinal or caudal segments of rats, and observed that the cartilaginous elements of such grafts grew after transplantation.

The modern literature on the behaviour of cartilage and of the articular ends of bone when transplanted is too extensive to be summarised. The movement which aims at transplantation of joints was initiated by Lexer in 1907. The total result of that movement has been to demonstrate that articular cartilage can be successfully transplanted—if it is still left *in situ* on the articular end of the bone. To secure this success, two conditions at least are necessary: (1) That in its new site the cartilage is enclosed within a synovial cavity—any part exposed to a non-synovial surface undergoes absorption; (2) that the graft is taken from the body of the "graftee." So sensitive is cartilage to circulating media that only those of the host meet its needs. The cartilage of the growth discs are even more sensitive to hormones and pressures; transplantation of a bone always damages the cells of the growth disc. Finally, there is the recent experience of Morestin.[1] He has utilised grafts of living costal cartilage to repair defects in the cranial wall and in the facial skeleton, and finds that such grafts hold and retain their health, life, and bulk for years after transplantation. It is very possible that, as in the case of bone, certain stresses are necessary to call forth the growth energy of cartilage cells. It is possible that the failure to obtain reproduction and growth in the cells of transplanted cartilage is due to the fact that such grafts are not subject to normal strains and stresses.

[1] "Les Transplantations Cartilagineuses dans la Chirurgie Reparatrice," *Bull. Acad. de méd.*, Paris, 1916, tome lxxv. p. 640.

CHAPTER XX

BONE-SETTING—ANCIENT AND MODERN

Owing to the present interest in "manipulative surgery"—the art of the bone-setter—I have been moved to examine the status and practice of representatives of the craft during the time of Cheselden, Hunter, Hilton, Paget, and of present-day surgeons. The results of my inquiry are given in the opening part of this chapter. I then trace the origin of our knowledge concerning internal derangements of the knee-joint—for it is disorders of the knee-joint which have provided and still provide "manipulative surgeons" with the chief part of their practice. Our knowledge of the condition and of the best means of treating it are shown to be discoveries of William Hey of Leeds, Thomas Annandale and Dr. W. Scott Lang of Edinburgh, and the host of surgeons of the present day who have had opportunities of studying a condition by methods which were not available in the days of their predecessors.

IN 1745, when Prince Charlie was unfurling the Jacobite flag in Scotland, a rebellion of a totally different character was afoot in London. The surgeons, who had been linked with the barbers for centuries, were petitioning Parliament to set them free from the "Masters or Governors of the Mystery and Commonalty of the Barbers and Surgeons of London." They were willing to sacrifice all claims to the worldly wealth of the old City Guild or Company if only Parliament would set them at liberty to work out a new ideal. Their prayer was granted, and we see the reason of their sacrifice and secession in the title given to the new body-corporate—"The Master, Governors, and Commonalty of the *Art* and *Science* of Surgeons of London"—which, in the opening years of the nineteenth century, became the Royal College of Surgeons of England. The new company was formed because the surgeons of London had made up their minds to have done with "Mystery"; henceforth there were to be no private, secret, occult practices, but a common fund of knowledge to which all might have access—a fund which every member might see, sample, sift, prove, and, if possible, increase. Nay, we have only to look at the record of a leading seceder —William Cheselden, the first of the modern line of English surgeons—to see that the aim of the new company, of which he was a first Warden, was much more than a pooling of personal experience. Arriving in London from Leicestershire in 1703, being then a lad of 15, he became apprenticed to Cowfeld the surgeon, and in 1711, at the age of 23, he was teaching anatomy on his own account, and writing the famous text-book which appeared two years later—in 1713—the first of many editions. Presently he was appointed surgeon to St. Thomas's Hospital, then in the Borough; introduced first the suprapubic operation for the removal of stone from the bladder; then he adopted the perineal route, and such was the dexterity of his hand and the accuracy of his knowledge of the anatomy of the part, that in 54 seconds he could perform, as a specimen in our Museum remains to show, the operation of lateral lithotomy. It was he who first devised a method of cutting a new pupil in the iris. Cheselden and his companions realised that, for the future welfare of their art, not only must surgeons pool and publish their experience, but it was also necessary, if surgery was to be regarded by learned men as a science, that they should found their practice on a complete knowledge of the structure and mechanism of the living human body.[1]

Thus before the middle of the eighteenth century was reached we find the apprenticed surgeons banding themselves to foster a knowledge of the art of healing. But there were also then, scattered through the country districts and towns of England and particularly of Wales, a class of practitioners, the legitimate heirs and sturdy representatives of the native British craft of orthopædic surgery known to the public as bone-setters; they were content to go on by rule of thumb, as their predecessors had done for many a century. The attitude

[1] William Cheselden died while at Bath in 1752 at the age of 64. In his later years he was surgeon to Chelsea Hospital, where John Hunter followed him as a student in 1749–50.

of the apprenticed surgeons to the rustic craftsmen we can learn from Cheselden himself. "Children," says he,[1] "are sometimes born with their feet turned inwards, so that the bottom of the foot is upwards. In this case the bones of the tarsus, like the vertebræ of the back in crooked persons, are fashioned to the deformity. The first knowledge I had of a cure of this disease was from Mr. Presgrove, a professional bone-setter, then living in Westminster. I recommended the patient to him, not knowing how to cure him myself. His way was by holding the foot as near the natural posture as he could, and then rolling it up with straps of sticking plaster, which he repeated from time to time until the limb was restored to a natural position. After that, having another case of this kind under my care, I thought of a much better bandage which I had learned from Mr. Cowper, a bone-setter at Leicester, who set and cured a fracture of my own cubit, when I was a boy." Thus we see that in Cheselden's day bone-setters were regarded as the legitimate practitioners of orthopædic surgery. A hundred years after Cheselden's time, as we have already seen, William John Little was handed over to a bone-setter to have his club-foot treated. There were treatises on bone-setting. There is in our College a copy of *The Compleat Bonesetter, wherein the Method of Curing broken Bones and Strains and Dislocated Joynts, together with Ruptures, vulgarly called Broken Bellyes, is fully demonstrated.* The author, however, was not a professional bone-setter, but a friar, Moulton by name, of the Order of St. Augustine, and the particular edition in our library is that which was "Revised, Englished, and Enlarged by Robert Turner" in 1656—Cromwell's time. We obtain a glimpse of the manner in which the ranks of bone-setters were recruited during the Commonwealth in the preface which Robert Turner wrote for this edition of Friar Moulton's book. He had not composed and made this work plain in the English tongue, he wrote, "to make Coblers cast away their Lasts and Awls, and such fellows, and straightway turn Doctors"; he would never write an English line on that account. The work, he assures

[1] *The Anatomy of the Human Body*, 6th ed., 1741, p. 57.

us, is intended "for the use of those Godly Ladies, and Gentlewomen, who are industrious for the improvement of their talent God has given them, in helping their poor sick neighbours."

In Cheselden's time, then, came the parting of the ways. The bone-setters continued to gather their knowledge in the school of local tradition, guarding their secrets jealously from all but their apprentices. They were content, like moles, to work in the dark without trying to discover why the means they employed sometimes succeeded and at other times failed. The apprenticed surgeons, on the other hand, set their affairs to ensure for their successors and for the public a commonwealth of knowledge; it became the bounden duty of each member of the new company to improve and increase the common fund by discovering the cause of disease, and by searching for a rational means whereby the cause might be removed or its effects combated. How far the Company of Surgeons succeeded in attaining that ideal is another matter.

Hunter followed Cheselden. He, too, had an intimate acquaintance of the empirics in Medicine. In Hunter's time Mrs. Mabb drove her carriage and four into London "to take charge of the dislocated limbs of the nobility and gentry," but the instance I cite, to illustrate Hunter's attitude towards empirical practice, is taken from his experience with a Mr. Plunkett, who had compounded a much-vaunted nostrum, containing arsenic, for the cure of cancer. "I was desired," Hunter told his students, "to meet Mr. Plunkett, to decide on the propriety of using his medicine in a particular case. I have no objection to meet anybody. It was the young one; the old one was dead, and might have died himself of a cancer, for aught I know. I asked him what he intended to do with his medicine. He said, 'To cure the patient.' 'Let me know what you mean by that,' I said. 'Do you mean to alter the diseased state of the parts? Or do you mean by your medicine to remove the parts diseased?' 'I mean to destroy them,' he replied. 'Well, then, that is nothing more than I or any other surgeon can do with less pain to the patient.'" Hunter then proceeds to tell how "poor Woollette, the engraver, died under one of those cancer-curers"—but there is no need to follow him further. The instance cited shows that Hunter, conscious that he could not cure cancer, was ready to meet and cross-examine the man who claimed he could stamp out that dire disease. On another occasion, Thurlow, the Lord Chancellor of England, asked Taylor of Whitworth, a rustic practitioner, to attend his brother, the Bishop of Durham, on account of a diseased condition of the rectum. Hunter was invited to meet the empiric, but was late. Taylor was asked to proceed without Hunter, but he refused, saying "he would do nothing till Jack Hunter came, for he had no opinion of any but him." Hunter having arrived, Taylor made his examination, and, declaring it to be a bad case, passed a candle besmeared with ointment. Hunter took up the box and asked what the ointment was made of. "That," said Taylor, turning to the Chancellor, "is not a fair question. No, no, Jack, I'll send you as much as you please, but I won't tell you what it's made of." The business man who has not fathomed the reason which induced medical men to drive "mystery" from their art and practice will condone the reticence of the rectal quack and condemn the surgeon's curiosity. And, indeed, in one small respect Hunter deserved that answer from Taylor, for on the opposite page to that from which the above quotation is taken[1] is a letter to his old pupil, Jenner, about the preparation of tartar emetic which Jenner had just discovered, and of which he proposed to publish a full description. "I would desire you to burn your book," Hunter told him, "for you will have all the world making it." There still remained something of the "old Adam" in Hunter, which is the more strange when we remember that he bestowed on surgery a bigger legacy of knowledge than all the men of his generation put together. Jenner, who was his pupil, would never have dreamed of keeping back the best he had to give his professional brethren, nor would Hunter's ardent disciple—Lister—have countenanced such an attitude. As time went on, the difference of motive between professional and quack becomes ever more marked, but even at the present time a

[1] Hunter's *Life*, by Ottley, Palmer's edition of Hunter's *Works*, vol. i. p. 87.

large proportion of educated men and women are still ignorant of the altruistic motives which have led to our medical knowledge becoming a professional trust for the public welfare.

We have obtained a glimpse of the status and practice of bone-setters in England during the eighteenth century through incidents culled from the lives of Cheselden and of Hunter. We are now to note their presence and practice in the nineteenth century. Hilton had experience of the bone-setter's practice of his time, and had not a good word to say for it. "Only the other day," he wrote,[1] "I had a handsome present sent to me from a young lady of a pair of crutches which I had lent to her some four or five years ago. She had then a severe disease of the hip-joint. I gave myself a great deal of trouble about her, and I believe I placed her joint in a comparatively healthy condition. Her note to me was—her compliments, and she sent back the crutches, having got well after nine months' treatment under a distinguished rubber at Brighton. She had been under my care for a considerable period, and no doubt she completely recovered by five months' further rest and quiet at Brighton." Hilton, you observe, refuses to credit the bone-setter with any share in the cure. He relates another case[2]—"the daughter of a grocer, suffering from severe hip-joint disease"—in whom he had succeeded in effecting a cure by bringing about an anchylosis of the joint. "I lost sight of the girl for some time, when the late Dr. Barlow said to me, 'I wish you would call at Mr. ——'s; his daughter is now very ill, and she wishes to see you.' I called, and as soon as she saw me she burst into tears and said, 'I am sure soon to die, and I am anxious to express to you my deep regret that I did not follow your advice. When I left your care I had a stiff hip-joint, and you advised me to be satisfied with that state. Following the advice of my friends, I went to Mr. ——, who said he could cure my stiff joint and make it movable.' This gentleman employed a great deal of force in order to distort and break down the bony union; fresh mischief was set up which resulted in large suppuration, so extensive as to resist all subsequent treat-

[1] *Rest and Pain*, 4th ed., p. 75. [2] *Ibid.*, p. 412.

ment." In that case Hilton rightly attributes disaster to the ignorance of the bone-setter.

In 1824, when Hilton commenced the study of Medicine at Guy's Hospital, Dr. John Shaw of the Windmill Street School was devoting his attention to the treatment of spinal deformities. At that time the care of bodily deformities, particularly those of the spine, was largely in the hands of bone-setters, and hence, in the course of his practice, Dr. Shaw became familiar with their successes and their failures. We have seen how he successfully unravelled the manner in which lateral curvature was produced, and how he searched for the best means of preventing and of curing the disorder. During that search he came to the deliberate conclusion that the results obtained in cases of lateral curvatures by bone-setters or rubbers[1] by means of friction and kneading and by movement and exercise were better than the results obtained by fully trained medical men who adopted the principle of rest—the principle which Hilton was destined to champion.

John Shaw was not blind to the self-deception with which the bone-setters of his time, as of every other period, were afflicted. "Now it happens," said he, "that some of the rubbers appear to do good by what seems to me to be mere parade, as when in finishing their operation they pretend, by certain manipulations, to put bones and muscles into their proper places. They pretend to push the bones of the spine back into place after the vertebræ have been 'softened to a jelly' by an hour's hard rubbing." Rubbing to soften and then the replacing of a bone which was never out of its normal position has been the standing practice of the British bone-setter for centuries, and has descended to their representatives in London to-day. "We have only to witness," Dr. Shaw wrote, "the operations of a professed rubber to be convinced that if the practice were superintended by one acquainted with anatomy and pathology, much benefit would result from it, while there would be no risk to the patient's life by its indiscriminate application." No one was more zealous in seeking for rational means of treatment than Dr. Shaw, yet

[1] Chapter XII.

in certain instances he believed that the means employed by men and women who were ignorant of anatomy and physiology were, when the full truth was known, not only more successful, but founded on a more rational method than that adopted by most qualified medical men. John Shaw, like John Hunter, never allowed pride or prejudice to stand between him and the acquisition of knowledge.

When the *British Medical Journal* appeared on New Year's Day, 1867, the pride of place was given to a lecture by Sir James Paget entitled "Cases which Bone-Setters Cure." In spite of Hilton's advocacy of rest as the most efficient means of treating injuries and diseases of bones, joints, and muscles—nay, it is possible just because of his indiscriminate advocacy of rest as a panacea for all forms of ailment—we learn from Sir James Paget, the acknowledged leader of his profession, that bone-setters not only were very much alive in his time, but actually prospered as the century went on. Bone-setters, we learn, sometimes came then, as they do now, rather unfairly by their successes. The surgeon who seeks the reward of skill by treating joint cases with rest until all inflammatory mischief has gone, waiting for the safe moment to come when the joint may be released from constraint, is in the position of the gardener who has nursed his choice pears to the point of perfect ripeness. A thief may break in and reap and possess the reward of all his labour. In the same way the bone-setter does reap, on many occasions, the reward of the surgeon's patience. But after making all allowances on such a score, Sir James Paget was convinced that there were a variety of conditions where the rubbings, the movements, the wrenchings of the bone-setter were more effective than the passive or neglectful attitude adopted by the qualified and registered medical man. Thus, in 1867, Sir James Paget succeeded in again compelling his professional brethren to examine the merits and demerits of the traditional British practice known as bone-setting.

At the time Paget delivered his lecture, Mr. Richard Hutton had established himself as a professional bone-setter in London, and his consulting-rooms in Wyndham Place, in the west end of London, were crowded with patients drawn from all grades of society—both men and women. He was the descendant of a family in the north of England, which, like the Taylors, Crowthers, and Masons, had exercised the art of bone-setting, from father to son, from time out of mind. He had been a successful upholsterer, and retiring from that trade devoted his leisure time to applying a traditional family skill to such neighbours as required and sought his help. From that humble beginning grew a reputation which brought him cases from all parts of the country, and necessitated the establishment of consulting-rooms in the west end of London. In the year 1865 he fell seriously ill, and was attended by Dr. Peter Hood, who was then a busy practitioner in Knightsbridge. Dr. Hood, who also had paid special attention to the treatment of sprains, and knew of Hutton's gratuitous services to the poor, refused to accept any fee for attendance, whereupon the bone-setter offered to impart to him all the secrets of his art. Dr. Hood accepted the gift—which the fashionable bone-setter cherished as the most valuable of his possessions—not for himself, but for his son Wharton, who had become a member of the Royal College of Surgeons some ten years before—in 1855—and was then assisting his father.

Hence it came about that Dr. Wharton Hood visited Hutton's rooms, and from day to day watched the bone-setter treat the kind of case which wearies the patience and exhausts the skill of the legitimate practitioner. Wharton Hood found the bone-setter to be a shrewd, observant man, practising his art in good faith and convinced that he could treat and cure certain disorders of joints which qualified medical men had attempted to treat and failed. For Hutton every disordered or stiff joint was "out"; every disorder was an unrecognised type of dislocation which had to be reduced by certain forced movements. "Don't bother me about anatomy," he would say tartly to Wharton Hood, when the latter would point out to him in a case that was to be operated on that the bones were all in their normal places. Seizing the patient's stiff knee, he would force it through certain predetermined movements until a "click" occurred, then, with a smile on his face, as he fixed the patient's eye, he would ask in his broad dialect, "Did you hear that?" Hearing was believing. "Now," said Hutton, "you're all right; use your limb." "A patient," says Wharton Hood, "who came hobbling down Hutton's passage with the aid of crutches, would, in a quarter of an hour, walk out briskly, would deposit his crutches in the carriage which had brought him, and walk home."

At the beginning of 1871 Hutton died, and after his death Dr. Wharton Hood[1] published a detailed account of the kind of cases he treated and of the methods he employed.[2] We there see, that in the latter part of the nineteenth century bone-setters no longer set fractured bones or reduced dislocations; that part of their work had been taken over by qualified medical men. All that remained to them was the aftermath of injuries or diseases of bones, joints, or muscles—lame or stiff joints which they treated by manipulation, rubbing, and movement. As a matter of fact, Hutton had no secrets to reveal; his methods were those which were known and used by his predecessors in Shaw's time and in Hunter's time. Nevertheless, Dr. Wharton Hood rendered medical men a service by making them certain of that fact.

We shall now examine a typical example of the kind of case which formed the staple of Mr. Hutton's practice—a condition of the knee-joint, which still provides the lineal successors of Mr. Hutton with their modern triumphs—nay, the kind of case which has led to the interposition of Parliament[3] on behalf of the professional bone-setter.

[1] Dr. Peter Hood died in 1890 at the age of 81; his son, Dr. Wharton P. Hood, in 1916 at the age of 82.

[2] *On Bone-Setting*, by Wharton P. Hood, M.R.C.S., M.D., 1871. *The Treatment of Injuries by Friction and Movement*, by Wharton P. Hood, 1902. *Modern Bone-Setting for the Medical Profession*, by Frank Romer, 1915.

[3] As a result of further representations by the Injured Soldiers' Parliamentary Committee, Mr. Peto, M.P., has been informed by the War Office that an Army Council Instruction has been issued to all commands in the United Kingdom directing that "no obstacle will be placed in the way of an officer or soldier who desires to avail himself of the services of a practitioner in manipulative surgery, who is not possessed of a medical qualification."—*Times*, 8th June 1918.

The case I am to cite is told by Dr. Wharton Hood as follows:

"Mr. A. sent for Mr. Hutton again. On his second visit I accompanied him, and what I witnessed made a great impression on my mind. We found the knee-joint enveloped in strapping; and when this was removed, the joint was seen to be much swollen and the skin shining and discoloured. The joint was immovable and very painful on the inner side. Mr. Hutton at once placed his thumb on a point over the lower edge of the internal condyle of the femur, and the patient shrank from the pressure and complained of great pain. 'What did I tell you two years ago?' asked Hutton. Mr. A. replied, 'You said my knee was out.' 'And I tell you so now,' was the rejoinder. 'Can you put it in?' said Mr. A. 'I can,' said Hutton, but declined to operate for a week. During the dialogue I carefully examined the limb, had satisfied myself there was no dislocation, and had arrived at the conclusion that rest and not movement was the treatment required. At the expiration of a week I went again to the house, Mr. Hutton arriving shortly afterwards. 'How's the knee?' was his inquiry. 'It feels easier,' said Mr. A. 'Been able to move it?' 'No.' 'Give it to me,' said Hutton, as he stood in front of his patient, who hesitated. 'Give me your leg, I say,' repeated the bone-setter, and the patient obeyed reluctantly. Mr. Hutton grasped the limb with both hands, round the calf, with the extended thumb of the left hand pressing on the painful spot on the inner side of the patient's right knee, and held the foot firmly by grasping the heel between his own knees. The patient was told to sit steadily in his chair, and at that moment I think he would have given a good deal to have regained control over his limb. Mr. Hutton inclined his knees towards his right, thus aiding in the movement of rotation which he impressed upon the legs with his hands. He maintained firm pressure with his thumb on the painful spot, and suddenly flexed the knee. The patient cried out with pain. Mr. Hutton lowered the limb and told him to stand up. He at once did so, and declared he could move the leg better, and that the previously painful spot was free from pain. He was ordered to take gentle exercise, and his recovery was rapid and complete." Thus Mr. A., after seeking help from the leading surgeons of London during a weary and suffering period of two years, was relieved by a turn of the hand given by a man who had never seen the inside of a knee-joint either in health or disease. We must be honest and admit—at least so far as London of 1876 is concerned—that medical men were as ignorant of internal derangements of the knee as bone-setters were then and are now. During the last fifty years, as we shall see, medical men have had every opportunity of mending that hole in their armour.

We have been confining our attention to London, but now, in our search for light and leading to help us in our task of unravelling the mysteries of bone-setting, we turn our eyes towards the North. We have seen that Owen Thomas came of a long line of bone-setters and had been initiated from his youth upwards in all the mysteries of manipulation, movement, and rubbing. And yet, after years of close and careful observation in a field which provided him with boundless opportunities, he was forced to the conclusion that the methods adopted and applied by bone-setters did much more harm than good. Sir James Paget's announcement in 1867 spurred him to further observation and comparison, and afterwards to use his pen to record his experience. He had the utmost respect for Paget's ability, but was fully conscious that by his opportunities, his training, and his experience he was better qualified than the London surgeon to pass an opinion as to the efficacy of the methods employed by bone-setters. We shall allow Owen Thomas to speak for himself:

"During the last twenty-six years," he wrote, "I have repeatedly tried manipulations to loosen joints crippled in their action, and have watched the practice of qualified and unqualified practitioners, famed for their skill as manipulators of diseased and injured joints; and again, I have been educated with a bias in favour of such treatment, yet, notwithstanding all this, unmistakable evidence of its evils has led me to discard it myself, and to advise others to avoid the adventurous treatment recommended by Sir James Paget."[1] "The irony of Fate," he wrote at another time, "that she should have selected me, specially trained, to oppose the teaching maintained in times past from the chairs of Esculapius in this town (Liverpool) by surgeons who are now enjoying their endless rest, and that she should so alter my preconceived opinions."

With Hutton's success in treating a case of internal derangement or "slipped cartilage" of the knee fresh in our minds, we turn to see how Owen Thomas dealt with this condition, which is particularly common in the north of England.

"The slipping of a cartilage," he says, "is another form of lesion in which bone-setters are thought to be adepts, but here again no special 'knack' is required.[2] This accident is so frequent that its relief is attained in most instances by moderate force in any direction that will give a little motion to the locked articulation. In many instances the procedure for relieving the disablement a patient learns for himself.

"Some five years ago I was consulted by a clerical gentleman, who stated that whilst getting out of bed the cartilage inside his knee had slipped out of place, and the sensation which this accident had caused remained. He also mentioned that this mishap had occurred to him before on more than one occasion, and that he had not, at any time, been properly relieved until he had submitted his limb to the manipulations of one of the metropolitan wrenchers. After this history of the case I examined the knee-joint and could find no trace of any physiological or mechanical derangement. The consulter, however, said he felt sure that there was some displaced cartilage, and as manipulation had eased him before, he wished it done again. His knee was firmly extended, rotated, and suddenly flexed, which was done rapidly and painlessly, for the patient, having no acute pain, relaxed all muscular control during our manipulation.

"After an absence of six weeks the reverend gentleman again presented himself, and, to my inquiry, 'How does the

[1] *Contributions*, Part II. p. 75, 1883. [2] *Ibid.*, p. 34.

joint feel?' he replied: 'After consulting you, a feeling of dissatisfaction induced me to hie off to the metropolitan practitioner whom I had consulted in my earlier difficulties, but his treatment has not given me ease.' I asked him to favour me with the details of the treatment he had undergone. The patient then informed me that, as soon as the specialist had examined his knee, he gave as his diagnosis that it was a case of displaced cartilage and that he would replace it immediately, which he attempted to do by manipulations like those I had performed; and as soon as the operator had finished, he said, 'Now you are all right; get up and walk,' which the patient did, but found the discomfort within the knee remained, and of this informed the practitioner, who repeated the manipulation, but with no success. The manipulations were several times repeated, but in vain, and the sufferer had returned to consult me again. On again examining the knee, there could now be detected a slight effusion of fluid within the joint, and my advice to him was the same as that I had given him during his first visit to me. This he consented to follow, and after a course of six weeks' treatment (fixation) the joint appeared normal and the patient was not sensible of any ailment."

Fifty years have passed since that clerical gentleman consulted Owen Thomas, and much has happened since then to throw light on the exact nature of his trouble. We now know for certain that he had reached a stage at which manipulations are powerless to set the cartilages of the knee right. For that increase of knowledge we are in no way indebted to bone-setters nor even to Owen Thomas. To trace the commencement of the movement which led us to an understanding of internal derangements of the knee-joint, we have to leave Liverpool and visit the busy city of Leeds—Leeds late in the eighteenth century—in Hunter's time.

We have seen how Charles White, the class-fellow of John Hunter, spent his energies in seeking to raise and rationalise the surgical practice of Manchester. The ideals of Cheselden's "Company of Surgeons of London" were leavening the cities of England. Leeds, too, had its Charles White—the one-eyed, but richly endowed William Hey. He was apprenticed to a surgeon in Leeds in 1750, when he was a mere boy of 14, and in 1757 entered St. George's Hospital, London, where Hunter had been house surgeon the year previous to his arrival. On returning to Leeds he took a leading part in founding its infirmary and in uniting his colleagues to form a medical society. The kind of man he was, the source from which he expected progress to come, we can judge from the fact that in 1801, when he was 65 years of age, he opened a school in Leeds for the dissection of the human body and the teaching of anatomy. At the beginning of his professional life he formed the habit of keeping written records of all his cases—a custom which he continued almost to his death in 1819 at the age of 83. In 1803 he culled from his notebooks a selection of cases which he published in book form under the title, *Practical Observations in Surgery, Illustrated with Cases.* One chapter of his book has a particular interest for us—a chapter to which he had given the heading, "On Internal Derangement of the Knee-Joint." Here we find the first clear recognition of a condition which has given bone-setters opportunities of scoring triumphs, not only during the hundred years which have come and gone since Hey laid down his scalpel, but also for many a century before he took it up. Hey describes four cases he had treated for the condition to which he gave the name of "internal derangement of the knee-joint." I shall select his second case to illustrate how far he had fathomed the nature of the trouble he had to deal with, and the success which attended the method of treatment which he adopted for its cure.

"In 1784," he wrote, "the Hon. Miss Harriet Ingram (now Mrs. Afton), as she was playing with a child, and making a considerable exertion in stretching herself forwards and stooping to take hold of the child, while she rested on one leg, brought on an immediate lameness in the knee-joint of that leg on which she stood. As the lameness did not diminish in the course of five or six days, I was desired to visit her.

"When I moved the affected knee by a gentle flexion and extension, my patient complained of no pain, yet she could not properly extend the leg in walking, nor bend it on raising the foot from the floor, but moved as if the joint had been stiff, limping very much and walking with pain.

"I thought it possible that the sudden exertion might, in some degree, have altered the situation of the cross ligaments within the joint or otherwise have displaced the condyle of the os femoris with respect to the semilunar cartilages, so that the condyles might meet with some resistance when the joint was put into action."

We see, then, that Hey hazarded the guess that the condition he had to remedy was somehow connected with the movable crescentic discs of cartilage which are interpolated between the femur and tibia at the knee-joint, but fortune never threw in his way an opportunity of ascertaining whether or not his guess was right. But, as we shall see, the treatment he applied was successful.

"To remedy this derangement I placed my patient on an elevated seat; I then extended the joint by the assistance of one hand placed just above the knee, while with the other I grasped the leg. During the continuance of the extension I suddenly moved the leg backwards that it might make as acute an angle with the thigh as possible. This operation I repeated once, and then desired the young lady to try how she could walk. Whatever may be thought of my theory, my practice proved successful, for she was immediately able to walk without lameness, and on the third day danced at a private ball without inconvenience." Although we cannot admire too much the manner in which he threw the quadriceps out of action by first pretending to extend the joint, meaning all the while to bend it, yet the sequel shows that Hey would have been wiser to have kept his patient in bed to permit a complete repair of the damage which had been done—for two years later he was again called upon to relieve his patient of a recurrence of the disorder.

My main purpose in writing is to bring instances of the kind just cited to the notice of my professional brethren, for I am convinced that if the medical practitioners of England had been familiar with the observations of William Hey and grasped his method of treatment, bone-setters would not be now flourishing in the west end of London, and there would have been no need for questions in Parliament concerning them.

We see, then, that recognition of "internal derangement of the knee" as a distinctive disorder was made in Leeds, but to understand how we came to know the nature of the lesion which gives rise to that disorder we have to follow certain "happenings" in Edinburgh during the latter half of the nineteenth century. In 1855 John Goodsir, the Master Anatomist, had become particularly interested in the mechanism of the knee-joint and gave special lectures on it.[1] It was during this episode in Goodsir's life that Owen Thomas came up from St. Asaph in Wales to study for his diploma, and close on his heels followed Thomas Annandale, son of a surgeon in Newcastle, who is presently to figure in this history. Goodsir tells us how his interest had again been roused in the mechanism of the knee-joint. His friend, Professor Meyer, of Zurich, had just published a book in which he drew the attention of medical men to a peculiar screw movement which occurred in the knee-joint when the leg was reaching the limits of full extension and when the various parts of the joint were being forced home. Goodsir saw that a screw movement occurred at the ankle and hip, as well as at the knee, but it was in the knee-joint that the screw device was most apparent. He recognised it as a very cunning mechanism for securing and locking a joint. There is one phase of the human step when the knee-joint must be rendered secure, and that is at the moment when one limb supports the whole weight of the body while the free limb is being swung forwards. The articular surfaces and ligaments are so arranged that the lower end of the femur is mechanically screwed against the opposing surfaces of the tibia, and thus the joint is locked and cannot again be opened until the screw is first undone. When the joint is locked, all its parts are taut and the semilunar cartilages are firmly compressed between the rims of the opposed articular surfaces of the femur and tibia. The fit of

[1] *The Anatomical Memoirs of John Goodsir*, edited by Wm. Turner, vol. ii. p. 220 et seq., 1868.

parts is so accurate—so exact—that the locking movement which marks the end of extension of the knee becomes impossible if there is the slightest degree of displacement of the semilunar cartilages or the interposition of a foreign body of even minute size. Goodsir paid particular attention to the semilunar cartilages; he observed that they retreated towards the popliteal space as the knee-joint was flexed; they were squeezed forwards as the locking movement occurred in complete extension, and that, in all positions of the joint, they helped to fill the inequalities which develop between the incongruous articular surfaces of the femur and tibia as the joint is flexed and extended. He made another important observation: he noted that Nature had taken the utmost precaution to prevent a vacuum being formed within the joint as the irregular contour of the femoral condyles plied on the flat articular surface of the tibia by planting a semi-fluid pad of fat, the infrapatellar pad, just where empty spaces would be developed. A similar provision has been made at every joint of the body, but it is particularly apparent at the knee, where the infrapatellar pad, covered by a synovial membrane, lubricates the articular surfaces as it rubs upon them. So fine is the fit, that from the side of the pads there are drawn out delicate folds which exactly fill the opposing articular crevices when the joint is fully extended. The semilunar cartilages, then, are parts of the mechanism for providing an exact fit between the articular surfaces, and giving complete security when the joint is locked.

Thomas Annandale, whose arrival in Edinburgh has already been noted, became a demonstrator of Anatomy to Goodsir, and in due time was appointed a surgeon to the Edinburgh Royal Infirmary. He was an eminently practical surgeon of the school of Syme, ready and accurate in his diagnosis, neat and expeditious with his knife, and when, in 1877, Lister vacated the chair of Clinical Surgery in the University on his migration to London, Annandale succeeded him. One day, early in November 1883, he admitted to his wards in the Infirmary a man suffering from "internal derangement of the knee-joint." He was a miner from the north of England, and his disorder—which is notoriously common amongst colliers —recurred so frequently that life became a burden and the earning of a livelihood an impossibility. "Having decided," so Mr. Annandale has reported,[1] "that the case was one of displaced semilunar cartilage and not likely to be cured by ordinary treatment, I, on 16th November 1883, performed this operation. An incision was made across the inner side of the knee . . . when it was seen that the internal semilunar cartilage was completely separated from its anterior attachment to the tibia and was displaced backwards about half an inch." The cartilage was drawn forwards, stitched into its proper place, the wound closed, and on 24th January the man returned to his work with perfect movement of the joint. In the following year he operated on three other cases of chronic derangement of the knee-joint. Thus a new phase in our knowledge of the condition which Hey was the first to differentiate from other troubles of the knee commences with this operation by Annandale, for he was the first to perform a deliberate and planned operation for the relief of a condition which had baffled the tricks of the bone-setter and the best means at the command of medical men. It is not the success which attended the operation which is our chief interest here, but the light which the introduction of this operation has thrown on the exact kind of damage which causes internal derangement of the knee-joint. Until surgeons commenced to operate for this condition we could only guess the nature of the damage; time showed that in the vast majority of cases it is due to a tear or rupture of a semilunar cartilage.

When Annandale was opening up this new field of surgical endeavour, and demonstrating, for the first time, the nature of the disaster known as internal derangement of the knee-joint, W. Scott Lang was completing his medical studies at Edinburgh University. He had reason to be interested in Annandale's discovery, for he himself suffered from the derangement, and two of his class-fellows, whose condition he studied, had also sustained the misfortune which so frequently overtakes athletic young men. Dr. Lang made his investiga-

[1] *Brit. Med. Journ.*, London, 1885, i. p. 779. See also 1887, i. p. 319.

tions and discoveries the subject of a thesis for the M.D. degree in 1886.[1] By experiments made on the cadaver, he discovered that on attempting to produce the lesion "pulling is of little use; it is the twist as does it." The formula is borrowed from Richard Hutton, the bone-setter, who found out by experience that if adhesions in a joint are to be broken down, it is useless to pull upon a limb; it must be twisted. He did not know that the muscles which surround a joint can safeguard it against pulling; but the knee-joint has no muscular protection to save it from being twisted. Hence in twisting the knee-joint there is no muscular opposition or resistance to be overcome. Dr. Lang discovered in the dissecting room that, when the foot is free, the semilunar cartilages of the knee-joint become stretched, torn, even ruptured and displaced, if the tibia is twisted on the femur; the same thing happens if the femur is twisted on the tibia, when the foot and leg are fixed on the ground. He found that the nature of the injury thus produced depends on the direction of the twist and the degree to which the knee is bent. If the knee is slightly bent and the upper end of the tibia twisted, so that the foot is turned outwards, or if the tibia is fixed and the femur rotated so that its internal condyle moves inwards and backwards, then the strain and damage fall upon the anterior part of the semilunar cartilage. If, on the other hand, the knee is much bent and rotation between the articular surfaces of the femur and tibia takes place in an opposite direction to that just described, then the stress and injury fall upon the external cartilage. He showed, further, that it is an easy matter, once the key has been discovered, to tell which of the two cartilages has suffered damage in every case which comes up for diagnosis. He pointed out as one of the diagnostic signs that the man who had suffered rupture of his internal cartilage walks with his foot turned in, for, in that position, the cartilage is unstrained and protected. The man whose external is damaged and displaced finds ease and safety in walking with his toes turned out.

[1] "Internal Derangements of the Knee-Joint," Edinburgh, 1886. See also *Edin. Med. Journ.*, Dec. 1886, Feb. March 1887.

Our tour in the North has shown us that the obscurity which enveloped the condition known as internal derangement of the knee-joint was dispelled by three discoveries: (1) the recognition of its existence and a definition of its nature by William Hey in 1784; (2) the demonstration of the nature of the injury by the operation introduced by Thomas Annandale in 1883; (3) the explanation of its production by Dr. Scott Lang in 1886. When we look round now to note the present state of our knowledge, we see that Annandale's pioneer operation has not only given relief to derelict knees which could not have been cured by any form of manipulative surgery,—whether applied by registered or unregistered practitioners,—but it has also given us a complete knowledge of the actual damage which such knees have sustained. Annandale's lead was followed at St. George's Hospital, London, first by the late Herbert W. Allingham, and afterwards by Sir William Bennett. In 1889, when Mr. Allingham[1] published the first book on the operative treatment of injured knee cartilages, he gave an account of nineteen cases which had been treated by operation up to that time, including four of his own. In 1900 Sir William Bennett[2] published an account of 253 cases which he had treated for internal derangement of the knee, and in fifty-three of which he had operated and thus seen the nature of the damage that had been done. In 1912 Mr. A. M. Martin[3] of Newcastle had operated on 509 cases, so that now, thirty-five years from the date of Annandale's first operation, it would be an easy matter to ascertain the exact nature of the damage sustained by the semilunar cartilages in 1500 cases of deranged knees. The important point for us is that in every case there is a tearing and rupturing of the substance of the cartilage or of its attachments with an effusion of blood to a greater or less extent into the breach thus caused. It is quite evident that the sooner the cartilage is replaced and its torn edges

[1] *The Treatment of Internal Derangement of the Knee-Joint by Operation*, London, 1889.

[2] *On the Use of Massage and Early Passive Movements in Recent Fractures*, London, 1900.

[3] *Brit. Med. Journ.*, London, 1912, No. 2, p. 1070.

brought into apposition by manipulation of the joint the better; but mere reduction will not heal the breach. That must be healed as any other wound—such as a cut on the back of a finger. In the case of a chap or cut of the latter kind, we notice that every time that we bend the finger the wound opens and is painful. If, however, we continue to use the finger the pain passes off. But we also notice that under such treatment the period of healing is prolonged, and the scar which results is extensive, whereas if we rest the finger the wound heals more quickly, and there remains merely the trace of a scar. There is no essential difference in the healing of a wound of a finger or of a semilunar cartilage; whether we give rest or give movement healing will take place, but it takes place quicker and better when we give rest.

We may justly claim that in modern times surgery has placed the treatment of internal derangements of the knee-joint on a scientific basis. Forty years ago we must admit that the trained medical man and the professional bone-setter were equally beset with the darkness of ignorance, but that is not so, or should not be so, to-day; yet when we turn to the number of the *Nineteenth Century and After* which appeared in October 1917, we find evidence to convince us that the lineal successors of the Mrs. Mabbs and Richard Huttons can still command a following, and can formulate a case for the recognition of their craft. When we examine the cases narrated by Mr. H. A. Barker in the *Nineteenth Century*, and the evidence he brings forward in support of the success of his treatment, we recognise he is dealing with cases in which a semilunar cartilage has sustained the kind of injury with which medical men are familiar, or should be familiar. The manipulations he applies are of the kind we have seen William Hey employ in 1784, and Richard Hutton in 1865—manipulations which are the common inheritance of medical man and bone-setter alike. But it is quite apparent, from the evidence brought forward by Mr. Barker—evidence which we have not any ground for rejecting—that there is a very considerable number of qualified men who know nothing of Hey, nothing of Annandale, nothing of the discoveries which their colleagues have made these forty years past—who have forgotten their anatomy and their physiology. So long as we have this class of medical men we shall have, and deserve to have, Mr. Barkers. When, however, that kind of medical man disappears from practice then the professional bone-setters will also cease to exist.

There is, in many forms of successful treatment, a psychological element, and there is no doubt that when manipulative methods are employed, whether by duly qualified or by unregistered practitioners, this element plays a part. It will be remembered, as I pointed out in another chapter, that when Duchenne applied his battery to the ears of deaf people, not to restore hearing, but to study the effects of exciting the facial nerve, patient after patient declared their hearing had been improved—which proved to be only a temporary and not a real improvement. In that instance the result was largely the result of a psychological influence. Dr. Marshall Hall quotes another example of this spurious element in successful treatment from the biography of Humphry Davy, the discoverer of nitrous oxide gas, who, it will be remembered, was at one time the apprentice of a medical man. The account of the incident is as follows:

"As soon as the powers of nitrous oxide were discovered, Dr. Beddoes at once concluded it must be a specific for paralysis. A patient was selected for trial, and the management of it was entrusted to Davy. Previous to the administration of the gas, he inserted a small thermometer under the tongue of the patient, to ascertain the temperature. The paralytic man, wholly ignorant of the process to which he was to submit, but deeply impressed, from the representations of Dr. Beddoes, with the certainty of its success, no sooner felt the thermometer between his teeth than he concluded the talisman was in full operation, and declared that he already experienced the effects of its influence. Davy desired his patient to renew his visit on the following day, when the ceremony was again performed, and it was repeated every succeeding day for a fortnight, the patient gradually improving during that period, no other application having been used than that of the thermometer."

Perhaps the example given by Mr. W. S. Penny of Bristol, in one of the earliest papers, to publish an account of the manner in which semilunar cartilages became displaced,[1] is more to the point. His father, a surgeon, had, on the advice of a duly qualified veterinary surgeon, turned a horse out to grass for three months to permit a wide open cut on the knee to heal up. The wound healed quickly and well, but at the end of the period the horse was found to be still seriously lame. Believing that the defect was permanent, the surgeon sold the horse for an "old song." A few days later the late owner was surprised to see his horse being driven along the road at a good pace and perfectly sound. "The new owner," it appears, "getting impatient of the horse's slow progress, gave it a smart cut with the whip. The animal, a well-bred, spirited beast, plunged forwards, and from that minute lost its lameness, and went as well as ever." In many cases it is the professional bone-setter's privilege to apply the sharp cut of the whip in cases of internally deranged knees and other forms of stiffened joints.

[1] *Brit. Med. Journ.*, London, 1888, No. 1, p. 1102.

INDEX